Concrete Box Girder Bridges

American Concrete Institute Monograph Series

Concrete Box Girder Bridges

Oris H. Degenkolb

PUBLISHED JOINTLY BY

THE IOWA STATE UNIVERSITY PRESS
AMES, IOWA

AMERICAN CONCRETE INSTITUTE
DETROIT, MICHIGAN

ACI Monograph No. 10

THIS MONOGRAPH is published in furtherance of ACI objectives in the fields of engineering education and technology. The Institute is not responsible for the statements or opinions expressed in its publications. Institute publications are not able to, nor intended to, supplant individual training, responsibility, or judgment of the user, or the supplier, of the information presented.

Library of Congress Cataloging in Publication Data
Degenkolb, Oris H   1918–
    Concrete box girder bridges.
    (ACI monograph; no. 10)
    Includes bibliographical references and index.
    1. Bridges, Box girder.   2.   Bridges, Concrete.
I. Title.   II. Series: American Concrete Institute monograph; no. 10.
TG362.D43   1977   624.4   76–43388
ISBN 0–8138–1815–X

Copyright 1977
American Concrete Institute
P.O. Box 19150, Redford Station
Detroit, Michigan 48219

First edition, 1977

Printed in the United States of America

Also in this series:

Lessons from Failures of Concrete Structures, by Jacob Feld—ACI Monograph No. 1

Evaluation of Concrete Properties from Sonic Tests, by E. A. Whitehurst—ACI Monograph No. 2

Freezing and Thawing of Concrete—Mechanisms and Control, by William A. Cordon—ACI Monograph No. 3

Durability of Concrete Construction, by Hubert Woods—ACI Monograph No. 4

Design of Flexural Members for Static and Blast Loading, by J. R. Allgood and G. R. Swihart—ACI Monograph No. 5

Hardened Concrete: Physical and Mechanical Aspects, by Adam M. Neville—ACI Monograph No. 6

Better Concrete Pavement Serviceability, by Edwin A. Finney—ACI Monograph No. 7

Precast Concrete: Handling and Erection, by Joseph J. Waddell—ACI Monograph No. 8

Testing Hardened Concrete: Nondestructive Methods, by V. M. Malhotra—ACI Monograph No. 9

ABOUT THE AUTHOR

ORIS H. DEGENKOLB, Civil and Structural engineer, Sacramento, California, has been involved in the design and development of box girder bridges in California since 1946. After obtaining the B.S. degree in civil engineering from the University of California, Berkeley, in 1942, he spent nearly four years as an engineering officer in the U.S. Navy. Most of this time was spent as Senior Repair Officer on board a battle damage repair ship in the South Pacific. He then worked as a structural engineering designer for several months, after which he joined the design section of the State of California Bridge Department. During more than thirty years in bridge design work he has been involved with all types of concrete, steel, and timber bridges; culverts; tunnels; pumphouses; buildings; retaining walls; bins; and miscellaneous highway related structures.

Although California had had a number of concrete box girder bridges designed and constructed before 1946, that year was the beginning of an accelerated highway program in which approximately 3100 reinforced concrete box girder bridges and 1100 prestressed concrete box girder bridges were designed and constructed by the California Bridge Department. Mr. Degenkolb was closely involved with the development of California's design specifications for box girder bridges throughout this entire period. Most of these specifications were later adopted by the American Association of Highway and Transportation Officials.

Contents

CONTENTS

1
Background

THE FIRST reinforced concrete box girder bridges were built in Europe and were generally long single spans with short cantilevers at each end. The cantilevers are primarily counterweights to produce negative moments at the supports and reduce the positive moments at midspan. The first reinforced concrete box girder bridge in the United States was constructed in 1937. Four states constructed box girder bridges before 1950, and their popularity increased. Twenty-six states had constructed them by 1960. This type of construction flourished in the West, primarily in California, where 3100 reinforced concrete box girder bridges and 1100 prestressed concrete box girder bridges have been designed and built.

When the first box girder bridges were designed, the available specifications for concrete bridges were for short spans, and the dead loads were a relatively small portion of the total design loads. Since the specifications required the same factor of safety for dead load as for live load, the usual types of concrete bridges were not economical for long spans because of the rapid increase in the ratio of dead load to total design load as the span lengths increased. The hollow girder concrete bridge was developed as a solution to the problem. The further use of continuity enhanced the economics of the box girder bridge.

Modified specifications for T-beam bridges were employed for de-

1

signing the first box girder bridges. Over the period of years the specifications have been modified on the basis of experience, judgment,
theory, and research. They have been continually revised to correct
deficiencies, achieve greater economy, and to keep up with new developments—of which prestressing is the most significant.

Practically all highway bridges built in the United States are designed according to the specifications of the American Association of
State Highway and Transportation Officials (AASHTO). Prior to
November 11, 1973, this organization was known as the American
Association of State Highway Officials (AASHO). These design
specifications have much in common with other building specifications. They are not a "cook book" that can be rigidly followed for
designing every bridge. They are not a substitute for experience,
good judgment, or imagination. They are generally suited for a
limited range of conditions which includes the most commonly used
span lengths, structure width, heights, and methods of construction.

The advent of prestressing increased the practical length for box
girder bridges and also permitted considerably thinner structures.
When highway safety standards in the United States discouraged the
use of columns on the motorists' right-hand side of highways in the
1960s the two-span continuous, cast-in-place, posttensioned, concrete
box girder became an ideal type of overcrossing structure. The economical depth-to-span ratio is considerably less than for other types
of construction. The difference in depth between prestressed concrete
spans and other types of construction is generally sufficient to provide for falsework over existing traffic, and the vertical clearance
under the completed structure is greater. If not built over traffic, the
roadway grades can be brought closer together, and the amount of
approach excavation or embankment is reduced considerably. The
completed structures are also aesthetically pleasing and require minimum maintenance. While approximately 90 percent of all reinforced concrete box girder spans so far constructed on state highways
in the United States are less than 100 ft in length, approximately 40
percent of California's 1100 prestressed concrete box girder bridges
have span lengths exceeding 150 ft.

Cast-in-place posttensioned concrete box girder bridges are now
economically competitive with other types of bridge construction
for spans from under 100 ft to over 300 ft (30–90 m). Cantilevered
concrete box girder bridges are competitive for longer spans.

In 1950 a prestressed box girder bridge was erected in Germany
by the cantilever method. That method of design and construction has extended the practical and economical span length of concrete box girders considerably. The longest completed span to date

is 787 ft (240 m), and construction of a 790-ft (241 m) span has been started. Longer spans may be expected in the future.

All engineered bridges have traditionally been designed by the Working Stress Design (WSD) method. In recent years there has been a concerted effort to switch to Load Factor Design (LFD) for all types of construction. The names Ultimate Strength Design (USD) and Load Factor Design are often used synonymously.

For the sake of economy each unit of a bridge should not be over-designed. If the actual live load placed on a bridge member is twice the design live load, and if the member was designed by the WSD method with 10 percent of the allowable stress assigned to dead load and 90 percent to live load, the member would be overstressed by 90 percent. If another member in the same bridge were designed by the same method, but 90 percent of the allowable stress is contributed by dead load and 10 percent by live load, a live load of ten times the design live load would be required on the bridge in order to overstress that member by 90 percent. The second member in this example would be grossly overdesigned since it would have a great reserve capacity remaining after the first member had failed com-pletely. All the members in a structure designed by LFD specifica-tions have nearly the same safety factor against failure by overloads. Maximum economy is realized because no part of the structure is un-necessarily overdesigned.

The U.S. Department of Commerce, Bureau of Public Roads, published *Strength and Serviceability Criteria—Reinforced Concrete Bridge Members—Ultimate Design* in August, 1966. A second edition was published in October, 1969, by the Federal Highway Administra-tion (successor to the Bureau of Public Roads). Both these LFD specifications were used on a limited basis as an alternative to the WSD AASHO specifications for highway bridges. A LFD specifica-tion for highway bridges was adopted by AASHTO in 1973, and it is anticipated that the American Railway Engineering Association (AREA) will adopt LFD specifications for railway bridges in the near future.

A highway bridge designed by LFD specifications has less main girder reinforcing steel and lighter girder stems than an identical bridge designed by WSD specifications.

2
Geometrics

2.1 SPANS

Reinforced concrete box girder bridges have been constructed with spans ranging from 30 to at least 235 ft (9–72 m). The general range for highway interchange structures is usually from 50 to 150 ft (15–46 m). Simple spans should be limited to a length of about 110 ft (34 m) because of excessive dead load deflections. Longer simple spans should be prestressed.

Prestressed box girder construction is suitable for spans of 60 ft (18 m) and over. Cast-in-place structures have been built with spans up to at least 460 ft (140 m). Segmental cantilever type construction has been used for constructing spans from 130 ft (40 m) or less to 787 ft (240 m).

There is a considerable overlap in the economics of reinforced versus prestressed box girder construction. All the structures in the ranges mentioned can be economically competitive under certain circumstances. For highway interchange construction either type may be suitable in the 80- to 120-ft (24–37 m) range with reinforced structures preferred for shorter spans and prestressed structures for longer spans. However, individual structures may vary from these limits.

Structures in a highway interchange are usually required to have minimum structure depths in order to keep approach roadwork, grades, and drainage problems to a minimum. Under these conditions

4

prestressed construction can be more economical, and also desirable for aesthetic reasons. The longer reinforced box spans may be
competitive with prestressed construction over rivers or roadway sep-
arations where there is an adequate amount of vertical clearance.
From an aesthetic point of view, the greater structure depth required
for a long-span reinforced box would be undesirable over a highway
that has minimum vertical clearance over the lower roadway. Thinner
structures are generally more pleasing in appearance.

Bridge span lengths in highway interchange structures are usually determined by highway geometrics and safety standards. The
bridge designer seldom has the opportunity to adjust them to the most
economical or structurally desirable proportions.

Structurally it is desirable to make the end spans approximately
3/4 of the adjacent interior span. In order to prevent uplift problems
at the abutments, end spans should not be much less than 1/3 of the
adjacent interior spans.

When span lengths are not dictated by highway geometrics or
other physical limitations, they are usually determined by economics
and aesthetics. Each bridge site is a special situation that must be
assessed on its own merits. Three-fourths end spans are generally economical and pleasing for most conditions. End spans that are one
half, or less, of the adjacent span should be used with caution and
usually are not aesthetically appealing.

2.2 STRUCTURE DEPTHS

To assure adequate stiffness to limit deflections that may ad-
versely affect the strength or serviceability of a reinforced concrete box
girder bridge, it is recommended that the depth-to-span ratios be approximately 0.060 for simple spans and 0.055 for continuous spans.
For prestressed concrete box girder bridges the corresponding ratios
should be about 0.045 and 0.040. For variable-depth structures of
either reinforced or prestressed concrete construction the midspan
depth of continuous bridges should be about 0.02–0.03 of the span
length, and the structure depth at support should be about 0.05–0.08
of the span length. These ratios should be considered as trial values
and may be varied to suit concrete strengths, required clearances,
aesthetic considerations, loading, and other pertinent factors. Although it is possible to design reinforced concrete structures with less
depth than with the ratios given above, such designs should be made
cautiously because of difficulties encountered with creep in extremely
shallow girders.

2.3 FRAMING

Long multispan bridges often require hinges in some of the intermediate spans in order to accommodate longitudinal movements due to temperature and shrinkage. Hinges have proven to be a source of trouble under seismic conditions unless adequate provisions are made to limit the distance they are permitted to open. Hinges generally do not provide resistance to flexing of the structure in a transverse direction. If hinges are located in two adjacent spans of a long structure, the portion of the structure between the hinges can rotate excessively without any appreciable restraint. This action can cause excessive strains and even failure of the entire structure during an earthquake. In order to prevent this condition it is recommended that hinges be separated by at least two bents.

2.4 GIRDER SPACING

The most economical girder stem spacing for ordinary box girder structures varies from about 8 to 12 ft (2.4–3.7 m). The average range for highway separation structures should be 8–9 ft (2.4–2.7 m) with greater spacings being used as span lengths and structure depths increase. Greater girder spacings require some increase in both deck and soffit thickness, but the cost of the additional concrete can be offset by decreasing the total number of girder stems. Fewer girder stems reduces the amount of formwork required, which results in a lower unit cost of concrete. It is not unusual for cantilever structures with very long spans to have girder spacings of 20 ft (6 m) or more.

The number of girder stems can be reduced by cantilevering the deck slab beyond the exterior girders. A deck overhang of approximately one-half the girder spacing generally gives very satisfactory results. This procedure usually results in a more aesthetic as well as a more economical bridge.

3
Decks

3.1 FUNCTIONS OF DECK

The deck of a concrete girder bridge has two primary functions: (a) to support the live load on the bridge and (b) to act as the top flange of the longitudinal girders. The concrete is designed to its full capacity in both directions. Ordinarily no consideration is made for the effects of maximum stresses acting in both directions at the same time.

3.2 DECK REINFORCEMENT

Decks ordinarily require four layers of reinforcing steel. Transverse reinforcing at the top and bottom of the slab is designed to support the live load and transfer the load to the main girders. Longitudinal "distribution bars" are laid on top of the lower transverse steel in the middle half of the deck span to aid in distributing the wheel loads to a greater longitudinal section of deck. Additional bars placed in the outer quarters of the slab span act as temperature and shrinkage reinforcement.

The main girder reinforcement is placed longitudinally just under the top layer of the transverse reinforcement. Temperature and shrinkage reinforcement is spliced to the ends of the girder rein-

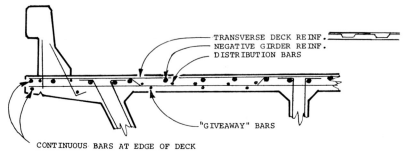

CONTINUOUS BARS AT EDGE OF DECK
Fig. 3.1—Typical deck reinforcement.

forcement in areas where the spacing of the girder reinforcement exceeds 18 in. (460 mm).

The four mats of steel require a considerable number of ties and supports in order to withstand the concentrated loads and forces encountered in the deck concrete placing operations.

It became a common practice for many bridge contractors in California to provide, at their own expense, longitudinal bars (usually #4s) directly under the bottom layer of transverse deck reinforcement. These "giveaway" bars support all the reinforcing steel in the deck slab. By using these additional bars the entire complex of reinforcing steel in the deck can be made into a rigid unit with fewer ties and a minimum number of supports. Since these bars have proven to be economically justified, the California Division of Highways now specifies them on contract plans. The "giveaway" bars in the middle half of the slab span are considered to be part of the distribution reinforcement (Figure 3.1).

Transverse deck reinforcement normally consists of continuous top and bottom bars alternating with "truss" bars that are bent so that they are at the top of the slab over the girders and at the bottom of the slab between the girders. The "truss" bars act as spacers between the top and bottom mats of steel and provide as well both positive and negative reinforcement. In variable-width structures, when girders are flared and the transverse deck span varies, it is costly to detail, fabricate, handle, and place the many differently dimensioned bars. Under these conditions it is generally more economical to specify short straight bars for positive and negative reinforcement. Under more extreme conditions it is sometimes more economical to use only continuous top and bottom bars. This is usually the case at the ends of skewed structures when the deck reinforcement is placed radially or normal to the girders and all deck reinforcing bars are different lengths.

Truss bars are relatively expensive to fabricate, but they simplify placement in the field. Straight bars are less costly to produce in the shop but require more labor to place. If truss bars are not used, steel chairs are required to separate the two layers of straight bars.

Although the use of truss bars in combination with continuous top and bottom bars is generally the more economical method of reinforcing decks, local labor conditions and practices might favor the use of all straight bars.

Deck reinforcement on curved bridges is generally placed radially with the spacing of bars measured along the centerline of the deck or some concentric reference line. This method requires a slightly greater amount of steel, but it reduces the great number of differently

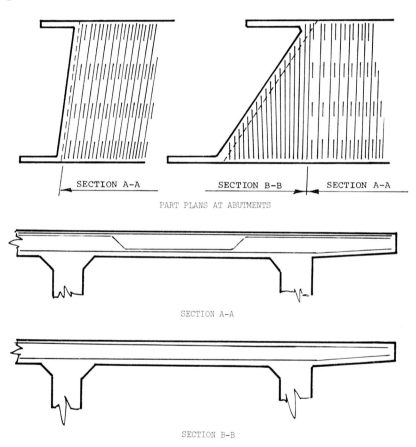

Fig. 3.2—Transverse deck reinforcement.

dimensioned truss bars and the additional expense of fabricating, labeling, handling, and placing the many custom dimensioned pieces required if the bars are placed parallel to each other.

It is generally preferable to place the transverse deck reinforcement parallel to the bents and abutments on skews up to 20 or 30 deg. The designer should be careful to check the additional area of steel required if it is not normal to the girders, and the plans should clearly note whether the bar spacing is measured along the centerline of the bridge or normal to the bars. For skews over 20 to 30 deg the deck reinforcement should be placed normal to the girders with only straight bars used in the remaining triangular end sections of the deck (Figure 3.2). Special details are required at the acute corners of the deck of a sharply skewed structure if there is a considerable deck overhang beyond the edge of the exterior girder to assure that the corners are adequately reinforced.

Continuous transverse deck steel could theoretically be considered to act in compression to reduce the calculated concrete stresses and decrease the deck thickness. For practical considerations this refinement is generally neglected.

Deck slabs are designed for one-way bending. When the deck adjoins an end diaphragm, or bent cap, or passes over an intermediate diaphragm, it is subjected to two-way bending for a short distance. Additional longitudinal reinforcement directly under the top transverse deck reinforcement, where there is no closely spaced negative girder reinforcement, prevents transverse cracking of the deck at these locations.

Deck reinforcement should be accurately placed and securely held in position by wiring at intersections spliced with #16-gage or heavier steel wire and by using precast mortar blocks, ferrous metal chairs, spacers, or other devices of sufficient strength to resist crushing or collapsing under applied loads. Bar supports that rest on the forms should not be made of materials that deteriorate, rust, shrink, or permit moisture to reach the reinforcement.

Although the California Division of Highways uses only Grade 60 reinforcement for bridge construction, it limits working stress of the steel to 20,000 psi (1400 kgf/cm²) in transversely reinforced deck slabs. This limit is based on experience in cracking of deck slabs.

3.3 MINIMUM COVERING

Since the under deck surface is well protected by the cell of the box, one-in. (25-mm) clearance to the lower reinforcing steel is adequate for all conditions. Coverage over the upper reinforcing steel should not be less than 1 1/2 in. (38 mm) for any condition. An additional 1/2 in. (13 mm) is recommended for marine environment. It is common practice to allow an additional 1/2 in. (13 mm) for wearing, which is not considered in calculating the effective depth of the slab. Special surfacings or protection should be considered if the deck will be subjected to the application of salt or other deicing chemicals that are harmful to concrete.

Extra coverage over the deck reinforcement is used by some states to allow for wear or the effects of severe weather conditions where the decks are subject to frost action and deicing chemicals. In some areas it is becoming a common practice to cut grooves in concrete roadways and bridge decks in order to increase resistance to skidding. Provision for this possibility can be included in the extra coverage allowed for wear.

3.4 DECK THICKNESS

Although most states use the minimum thickness of deck required by design requirements, there is a considerable variation, and a number use thicknesses that are considerably greater. For similar girder spacing the thickness of decks in various states varies from 6 to 9 in. (150–230 mm).

Over a period of years there has been a trend to increase concrete and steel working stresses, which has resulted in more flexible decks. This, in conjunction with increasing wheel loads, is often considered to be one of the factors that contribute to deck deterioration. There are a number of very flexible, badly cracked bridge decks that are performing very satisfactorily and will probably continue to do so for many more years. However, they are in very mild climates where there are no other adverse factors.

3.5 DECK FORMS

The forms for box girder bridge decks are usually supported by timber ledgers bolted near the tops of the newly completed concrete girder stems. Removing the forms, or "lost deck," is difficult and expensive; so they are generally left in place. Consequently, it is common to use used materials—quite often the same forms that were used for the stems. If identical fillets are used at the top and bottom of the girder stem, the bevelled edge of the girder form, which formed the bottom fillet, can be used again for forming the upper fillet.

Thin concrete slabs, either reinforced or prestressed, which support the fresh deck concrete and become an integral part of the bridge deck, have been used quite successfully. Sheet metal and plastic deck forms that remain in place have also been used. They are usually corrugated in order to obtain the required stiffness with a minimum amount of material, for the sake of economy. The corrugations require a small additional amount of deck concrete. If steel forms are used, consideration should be given to the possibility of eventual rusting that will stain the concrete.

3.6 DECK DETERIORATION

Concrete box girder bridges are subject to the same deck deterioration problems as any other type of bridge having a concrete deck. Whether or not a deteriorated deck on a concrete box girder bridge is any more serious than a deteriorated deck on a T-beam, concrete slab, or steel girder bridge is more an academic than a practical problem.

The primary cause of deck deterioration is improper construction that results in vulnerability to frost action or corrosion. When a deck is vulnerable to corrosion, corrosion is accelerated if salts are present. Most salt on bridge decks comes from deicing or ice prevention operations. Bridges near the ocean that are subjected to direct salt water spray, or spray carried by offshore winds, are also subject to deterioration. Salts in contact with reinforcing steel cause rusting, which leads to undersurface fracturing and potholing of the concrete. Salts generally get to the reinforcing steel through porous or cracked concrete.

The best methods of prevention against such deterioration are to build corrosion resistant decks or to eliminate the prime factor—salt. However, the latter is beyond the influence of the designer or the contractor.

If the concrete surface will be exposed to salts, additional cover over the steel should be specified, and the concrete should be properly proportioned, placed, consolidated, and cured. Deterioration occurs much faster if the concrete is porous and has cracks. The greater the number of cracks of sufficient width, the more rapid will be the deterioration.

Extra concrete cover over the reinforcing steel increases the time that it takes the salt to reach it. However, when the salt does reach the steel the expansive action of the corroding steel will cause popouts and spalling. Frost action of water in the cracks and heavy axle loads tend to widen and deepen the cracks, which accelerates the deterioration process.

In practice, a high-quality, crack-free concrete deck that will prevent salts from reaching the reinforcing steel is seldom obtained. This has led to efforts to seal the deck surface. Many methods have been tried from the simple application of linseed oil to various combinations of membranes and surfacings. Although some methods are showing promise of being successful, none is yet generally accepted as the ideal solution to the problem.

Until a few years ago it was almost impossible to detect a corrosion problem in decks with overlays until the problem was serious. Half-cell corrosion detection devices are now available which can show if there is a corrosion problem in a deck even though it cannot be detected visually.[1] Some waterproof membranes have been used successfully for a number of years, but there is no assurance that occasional leakage will not occur. Many authorities agree that continued use of overlay systems is justified and will prevent all but minor damage to bridge decks where salts are present. Membrane leakage is most commonly due to blow holes and blisters; inadequate sealing at joints, edges, and curbs; and membrane puncturing by construction activity. These defects can be overcome by proper workmanship and attention to details, although this often is difficult at the job site.

A method of cathodic protection, in which an electrical current is passed through the deck from the reinforcing steel to a sacrificial metal, has been in use for a number of years and appears to be promising.

The proper use of entrained air in properly proportioned mixtures prevents damage to concrete by frost action. The use of such concrete in structures of proper design, especially with regard to concrete cover over steel, prevents corrosion. A minimum cement content of 700 lb/yd³ (415 kg/m³) of concrete is also recommended.

On a group of bridges that had severely deteriorated decks aggravated by heavy salting, it was noted that the worst decks had few wire ties connecting the reinforcing bars in the top mat. Some of the bars were separated vertically by as much as an inch (25 mm) of concrete. Decks that had the reinforcement tied more securely were not deteriorated as severely. It appeared that the loosely placed deck steel may have moved during deck placing operations and formed voids around the steel. The voids permitted the salt to rapidly contact a greater area of steel, once it penetrated the cover, and hastened the deterioration of the deck. Although this phenomenon has not been investigated thoroughly, it seems that it might be a contributing factor in accelerating the deterioration process in some instances. The deck reinforcement should be securely tied together for other reasons as well.

It is ordinarily assumed that when a live load deflects a deck slab, the slab returns to its original position and condition. After a great number of flexures it is probable that a grain of sand is displaced in a minute crack and, after continued working, causes other grains to become dislodged. As cracks enlarge, water, salts, and other deteriorating agents are free to attack the reinforcing steel and faces of the cracks. There are many cracked bridge decks in areas with mild climates that will undoubtedly give many more years of satisfactory service. The same bridge decks located in areas where salts are used for deicing would unquestionably last only a short time.

3.7 CONSTRUCTION

The need for good quality bridge decks cannot be overemphasized, especially in corrosive environments and where roadways are subject to deicing chemicals. Each phase of bridge deck construction—design, quality of materials, proportioning, mixing, placing, consolidation, finishing, and curing—is important, and any one of these factors, if not done properly, can result in an unsatisfactory deck. A comprehensive discussion of all phases of concrete bridge deck construction is given in the ACI standard, "Recommended Practice for Concrete Highway Bridge Deck Construction (ACI 345–74)." by ACI Committee 345.

4
Soffit

4.1 FUNCTION

The soffit of a box girder bridge functions as a compression flange for negative girder moments; it contains the positive girder reinforcement and is also a significant architectural feature. The soffit slab permits a box girder bridge to be considerably thinner than a T-beam bridge with the same span and also makes much greater span lengths practical.

4.2 THICKNESS

The first box girder bridges had no criteria for designing the soffit slab. Some designers used a minimum thickness of 5 1/2 in. (140 mm) required for transverse #4 bars above and below longitudinal #11 main girder reinforcement with 1 1/2 in. (38 mm) of cover under the bottom steel. The cover over the top soffit reinforcement, usually 1 in. (25 mm), can be less than the bottom because it has better protection.

As box girder structures became more common and greater span lengths were used, girder spacings were increased to achieve greater economy. Although horizonal shear stresses between the girder stem and soffit slab were within reasonable limits, there was a feeling among some engineers that the soffit should be thickened for

15

Fig. 4.1—Construction of soffit forms.

greater spacings. It was arbitrarily determined that the soffit thickness should be 5 1/2 in. (140 mm) minimum, 1/16 of the clear distance between girders, or thicker if required for design. This specification was later modified so that the soffit slab need not be thicker than the deck slab, unless required for other reasons.

The thickness of a soffit slab is usually kept to a minimum at midspan and is thickened as required at the supports of continuous spans in order to keep compressive stresses, produced by the negative girder moments, at a desirable level.

Figure 4.1 shows the placing of the joists on which the plywood forms for the soffit will be placed.

4.3 FORMING

On structures with a constant deck cross slope it is logical to make the soffit parallel to the deck so that all girder stems are the same depth. When the roadway of a box girder bridge is crowned or has a change in the transverse cross slope, there is usually an option: to make the soffit parallel to the deck or to make it straight between the two exterior girders. The first option makes all the girders the same depth, the stirrups in all the girders identical, and requires a minimum amount of concrete. It has the disadvantage of requiring a break in the soffit forms, which may be troublesome in very narrow structures but which becomes less of a problem in wide structures. The second option simplifies the construction of the soffit forms but requires different depths of girders, different lengths of stirrups in each of the girders, and a greater amount of concrete.

For either option, the design of the structure is the same and should be based on the minimum girder depth. The variations in girder depths, dead load, reinforcing steel, prestressing force, and other factors are usually negligible from a practical point of view. If the deck is crowned, the soffit may be either straight or crowned.

In the interest of economy, it is suggested that the contractor be given the choice of using whichever option he prefers. However, the method of payment should be clearly defined so that the cost to the owner is the same regardless which option he chooses.

The underside of the soffit slab should be finished by filling the the holes or depressions in the surface of the concrete, repairing all rock pockets, removing fins, and removing stains and discolorations where they can be seen by people if they regularly pass under the bridge.

4.4 VENTS

Vent holes should be placed in the soffit slab in each cell for draining curing water during construction and any rain water that leaks through the deck. Five-in.-diameter holes have proven adequate. It is recommended that two vents per span in each cell be used. One vent should be located at the low point and the other at the opposite end of the cell. More or larger vents should be considered for cantilever or other unusual type structures in which temperature differentials might be a controlling design factor.

4.5 LIGHTS

In cities, lights are frequently desired under bridges in order to provide lighting for pedestrians and motorists under the structure. This is particularly necessary on very wide structures with minimum vertical clearance. Recessed lighting fixtures can be cast directly in the soffit slab in order to minimize encroachment on vertical clearance under the structure.

5
Girder Stems

5.1 GENERAL

Girder stem widths are determined by design, construction, and economic considerations. They resist vertical shear and usually a relatively small proportion of the girder moments; so they are consequently much thinner than T-beam stems. T-beam stems must be proportioned to resist negative girder moments; to contain all the positive girder reinforcement; and to resist shear.

AASHO bridge design specifications originally permitted exterior girders to have considerably less shear carrying capacity than interior girders. When it was observed that many of the exterior girders designed by these specifications developed diagonal shear cracks near the supports, the specifications were revised to require that the exterior girders should in no case have less carrying capacity than an interior girder.

A reinforced concrete box girder bridge designed for an HS 20–44 live load by the present Load Factor Design specifications has girder stems with somewhat less shear carrying capacity than a similar bridge designed for the same live load by the Working Strength Design specifications.

Design specifications do not require any minimum girder stem width as long as allowable stresses are not exceeded. Girders which are of minimum thickness for design considerations require less concrete and minimize the dead load, but they may be costly to build because of problems encountered in placing the concrete. Thick stems may reduce the labor cost per unit volume for placing but require heavier falsework and more concrete. The additional concrete is costly and increases the dead load, which in turn increases the quantities of concrete and reinforcing steel in the rest of the structure. The optimum girder stem width is thus a compromise between the cost of materials, cost of labor, and the capabilities of available construction personnel.

It is usually necessary to flare the stems of box girders at the supports in order to resist vertical shears. Stems must also be flared at the ends of prestressed structures to accommodate prestressing anchorages.

In the United States there is no standardized box girder stem width. The various states use minimum widths varying from 8 to 12 in. (200–300 mm) for reinforced box girders; 8 in. (200 mm) is generally considered to be a minimum for placing concrete with two mats of steel. Prestressed box girders generally require a minimum thickness of 12 in. (300 mm) in order to accommodate the larger prestressing ducts. Stems less than 12 in. (300 mm) thick can possibly be used if the structure is designed for one specific method of prestressing.

Consideration should be given to increasing the stem widths of unusually deep girders because of the difficulty in placing the concrete. Some designers have used the rule of thumb of 1 in. (25 mm) for each foot (300 mm) of height, with an 8-in. (200-mm) minimum.

In cantilever structures thermal differentials and transversely prestressed decks can cause problems if girder stems are thicker than necessary. In these instances it is desirable to keep the girder stems as thin and as flexible as possible.

5.2 CONSTRUCTION

Some states require, or permit, the soffit slab to be completed first and the stems placed later. Other states require that the soffit and stems be placed without a construction joint. In the latter case, concrete is placed in the stem forms as soon as the soffit concrete has hardened sufficiently to resist the fluid forces of the fresh stem concrete. There is a tendency for the stem concrete to hang up on the longitudinal web reinforcement and for voids to form under the horizontal bars. The stem concrete should be adequately compacted by vibration to assure that it flows around and under the horizontal reinforcement. If the soffit concrete has not set up sufficiently to resist the pressures, the stem concrete will flow into the soffit and, if

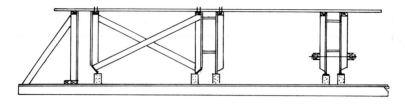

SOFFIT AND GIRDER STEM FORMS

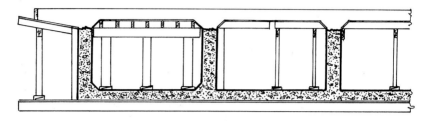

DECK FORMS

Fig. 5.1—Typical box girder forms.

this happens after the consolidation has stopped, voids will be formed under the horizontal bars.

All of the reinforcing steel (and prestressing ducts if the structure is prestressed) in the soffit slab and girder stems is generally placed and tied after the forms for the soffit slab and outer faces of the exterior girders are completed. Forms for girder stems are then placed on concrete blocks, which provide support, act as spacers, and become a part of the soffit slab. The method of forming fillets between the soffit slab and stem is illustrated in the exterior cell of Figure 5.1. The interior cell is shown without the bottom fillets. All cells are ordinarily constructed either with or without bottom fillets.

When the stem forms are to be used on a number of structures that have different structure depths, it is possible that the forms will be much deeper than required for some of the structures. Concrete, then, will not be placed the full depth of the forms. In some instances the forms interfere with the top hooks of the stirrups, in which case the top hooks can be bent in place after the forms are removed. Different methods of forming the stems are constantly being tried in an effort to reduce costs.

One of the economic advantages of box girder bridges is that only the outer faces of the exterior girders need to be finished. Specially treated forms for the outer surfaces of box girders have been used to obtain a smooth high grade surface that does not require additional finishing. The results were not completely successful. Although most of the surfaces were very satisfactory, some blemishes and minor rock pockets required patching and refinishing. It was impossible to make the necessary repairs match the surrounding areas. Also, vibrators blemished the forms during placing operations, and the repaired blemishes did not give a uniform finish when the forms were reused.

Unless special architectural treatments are required, the outer surfaces of the exterior girders should be finished by filling all holes or depressions in the concrete, repairing all rock pockets, and removing fins, stains, and discolorations. If the surface is not then smooth, even, and of uniform texture and appearance, it should be sanded with power sanders or other abrasive means until a satisfactory finish is obtained. Bulges and other imperfections should be removed by using carborundum stones or disks.

Figure 5.2 shows the plywood forms for the soffit in place and the outer face of the exterior girder being constructed. The vertical

Fig. 5.2—Forms for soffit and outer face of exterior girder.

Fig. 5.4—Construction of a continuous span posttensioned box girder bridge.

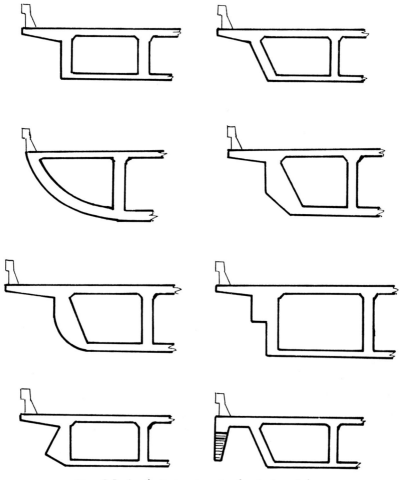

Fig. 5.5—Aesthetic treatments of exterior girders.

kgf/cm^2), for designing stirrups in order to control cracking in the girder stems. In Load Factor Designs it is assumed that $f_y = 50,000$ psi (3500 kgf/cm^2), rather than $f_y = 60,000$ psi (4200 kgf/cm^2), for the design of stirrup steel for the same reason.

5.6 ARCHITECTURAL TREATMENT

The exterior girders of concrete box girder bridges are much more adaptable to special architectural treatments than any other type of bridge girder. Figure 5.5 illustrates some of the exterior girder configurations that have been used quite extensively. The appearance of the plain vertical or sloping faces can be very pleasing. The most frequent reason for using other configurations is to make the super-structure appear as thin as possible. However, the entire structure should be designed to harmonize with its surroundings, and all parts of the structure—girders, railing, columns, abutments, wingwalls, and so forth—should be compatible with each other.

The outer concrete surfaces are normally finished to produce a smooth even surface of uniform texture and appearance. The de-

Fig. 5.6—Mural formed in exterior face of box girder bridge using different thicknesses of plywood.

gree of care in building the forms and the quality of the materials have much to do with the amount of finishing work required and the final appearance of the work. Unsightly bulges or other imperfections should be ground down, if necessary, and the entire surface finished to give a uniform texture and appearance (see Section 5.2).

In addition to varying the shapes of the exterior surfaces, various other methods are used for obtaining architectural effects. Exterior faces are sandblasted to give a dull, uniform appearance or are stained and painted various colors. White cement is used for arch-shaped curtain walls suspended from the deck overhang, which contrasts with the normal gray concrete of the exterior girder. Form panel joints are accented to give a uniform block pattern. On one job, five thicknesses of plywood with suitably cut holes in each were used for forming a mural on the exterior girders (Figure 5.6). Commercially available form liners give a wide variety of standard surface patterns and can be custom made.

6
Diaphragms

6.1 END DIAPHRAGMS

End diaphragms are generally used between the girder stems at abutments and piers on all concrete box girder bridges. If the framing of the bridge is such that no movement must be accommodated at the ends, it is not necessary to use a separate abutment structure (Figure 6.1). The end diaphragm can be placed directly on piles or a spread footing and also used for retaining the approach roadway fill. This has proven to be a very successful and economical detail. When separate abutment structures are used, the end diaphragm transfers the reactions of the girder stems to the abutment bearings.

Pier or bent caps that are within the limits of the deck and soffit slab act as an unusually effective and rigid diaphragm. In some types of construction in which bearings are used under the soffit slab at the bents, some sort of a diaphragm, either partial or whole, is required to transfer the girder reactions to the bearings. Diaphragms at abutments and bents provide transverse stability to the girders.

End diaphragms in skewed structures tend to increase the difference between the bearing reactions and may even produce uplift at heavily skewed abutments.[2] They are also subjected to high bending and torsional stresses. Even though these tendencies are

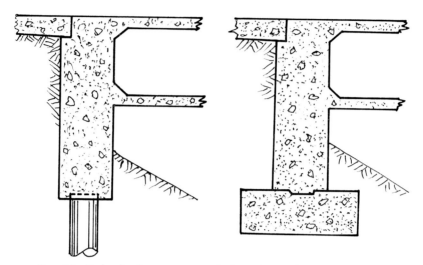

Fig. 6.1—End diaphragms extended and used as abutments.

undesirable, they are seldom serious enough to cause a great amount of concern.

Deck slabs are ordinarily designed for one-way bending. Since a diaphragm supports the deck and introduces a negative longitudinal moment in the deck, longitudinal reinforcement should be placed under the top of the deck slab to prevent transverse cracking.

Some very large box girder bridges carry highway traffic on the top deck and railroad traffic inside the box cells.[3,4] With these examples it is obvious that box girder structures can be designed and constructed successfully without either end or intermediate diaphragms.

6.2 INTERMEDIATE DIAPHRAGMS

The usefulness and action of intermediate diaphragms in concrete girder bridges seems to have been misunderstood and overemphasized in the past. Recent tests and theoretical analyses indicate that they have very little benefit in tangent bridges. Although there is a widespread feeling that intermediate diaphragms are desirable or necessary in sharply curved concrete box girder bridges, there is very little quantitative information available on that subject. In some structures diaphragms can actually be more harmful than beneficial.

Intermediate diaphragms are useful when a single load is placed on a box girder bridge and the load is applied at the diaphragm.

This condition of loading is not realized under usual conditions. To be fully effective for loadings that are ordinarily applied to highway and railroad bridges, numerous diaphragms would be required at very close spacing. It also might be desirable, theoretically, to use diaphragms in a structure that has very stiff deck and soffit slabs and very light girder stems, but this combination is not realized in practical designs.[5]

Prior to 1969, AASHO bridge design specifications required intermediate diaphragms to be placed at a maximum spacing of 40 ft (12 m). In 1969 the spacing was increased to 60 ft (18 m) for tangent structures, and curved girders were to be given special consideration. The California Division of Highways has used a spacing of 80 ft (24 m) in tangent structures since 1957 and has omitted them from many tangent structures since 1973 and has not observed any adverse effects.

Intermediate diaphragms cause the deck to act as a two-way slab, but, unless the spacing of the diaphragms is nearly the same as the girder spacing, any advantage of two-way bending in the deck is lost. Such close spacing of the diaphragms to benefit the deck design is not economical. When intermediate diaphragms are used, longitudinal reinforcement should be placed near the top of the deck slab over the diaphragms to accommodate the negative moments in the deck slab caused by the diaphragm.

The advantages of using intermediate diaphragms in tangent structures are very minimal. It is generally concluded that the few advantages they might produce are not worth the trouble and expense of providing them.[2,6,7]

7
Construction Joints

7.1 GENERAL

Construction joints are a practical necessity to keep the total quantity of concrete placed in any working day to a reasonable amount, to permit all necessary operations to proceed without undue interference, and, in the case of long multispan structures, to permit the reuse of falsework and forms.

7.2 PLACING SEQUENCES

The first continuous box girder bridges were constructed in sections with construction joints located near the dead-load inflection points. The midspan soffit slabs and stems were completed first and the connecting sections over the bents placed second. The deck slab was placed last in either one continuous operation or with the section over the piers completed at a later time. The reasoning behind this procedure was that if the midspan sections were placed last, settling of the falsework footings, compressive shortening, and take-up of the falsework joints would induce premature moments and crack the fresh concrete over the bents.

As box girders became popular for long viaduct structures, these limitations increased the construction time considerably and required contractors to tie up considerable amounts of falsework and forms for long periods of time. With the experience of constructing many structures by many contractors at widely varying site conditions, it was determined that the placing sequence could be revised considerably.

If good quality falsework is used and the falsework footings are adequate, it has been found satisfactory to permit the contractor to start at one end and place the soffit slab and stems continuously for the entire length of the bridge or to a construction joint at approximately the 0.2 point of the span from any pier. The next placement continues to the end of the bridge or to another 0.2 point of a span. Interim reinforcement (see Section 10.3) should be placed in the tops of the girder stems to control cracking. The deck can then be placed at a later date in the same fashion.

7.3 HORIZONTAL CONSTRUCTION JOINTS

It is obvious that it would be undesirable or difficult, for practical considerations, to place the soffit, stem, and deck concrete in a single operation. Some type of longitudinal construction joint is a practical necessity in order to permit the deck forms and reinforcing steel to be placed after the soffit concrete has been placed.

Some specifications require that the soffit, stems, and deck be placed separately. Other specifications require that the soffit and stems be placed at the same time and the deck at a later date.

The horizontal construction joint at the top of the stem is logical and desirable for a number of reasons. The side forms for the stems can be easily removed and reused for forming other girders or the deck slab; the completed girder stems are sturdy supports for the deck forms, which must resist the forces involved in constructing the deck; and, if the stems and deck are placed at the same time, a large amount of accumulated debris would have to be removed from the deep, narrow girder stems, which could be a considerable job—especially with stirrups, longitudinal reinforcement, and large prestressing ducts inside the forms. Also, stem concrete settles appreciably after it is placed, and if the deck concrete is placed too soon, a horizontal separation will be formed at the junction of the two units.

Conclusions based on the results of laboratory tests have led to a considerable change in thinking concerning the use of construction joints. It was once believed that shear keys were required on a horizontal joint in order to get a good interaction between two layers of concrete. These shear keys were made by various means and were often of obviously questionable value. Tests indicate that keys used with a rough bonded contact surface do not change the strength of the connection. The slip movements required to develop the keys are greater than the movements for a bonded surface. It is important that the contact surface be left rough and no attempts be made to smooth the aggregate into the paste. Depressions and peaks should be approximately 3/8 in. (10 mm) below and above the average level.[8] Blast cleaning is the most commonly used effective method for cleaning construction joints. Stirrups and other reinforcing steel passing through the joint increase the resistance to shear on the interface.

7.4 VERTICAL CONSTRUCTION JOINTS

Although shear keys are not the most efficient type of horizontal construction joint, they are still unequaled for vertical joints. This is due to two basic factors: *(a)* the method of constructing the joint and *(b)* the fact that hardened concrete shrinks as it dries and drops in temperature.

The surface of a horizontal joint, if it is not overfinished, is rough and each minute bump or depression is a natural key. When the second layer of concrete is placed on a horizontal joint, shrinkage may later produce numerous vertical cracks normal to the surface, but there is good bonding on the interface if the surface has been properly cleaned by blast cleaning or other equivalent means.

Concrete is placed against forms to form vertical construction joints, and the finished surfaces are smooth. A considerable amount of blast cleaning or hammering would be required to provide the same rough surface that can be obtained on a horizontal joint with relatively little effort. Some time after concrete is placed against a vertical joint, shrinkage tends to produce a fine crack at the joint, and the amount of bond between the old and new surfaces, if any, is questionable. For these reasons vertical joints in members that carry high shear force are generally made with offset steps so that the horizontal projections act as bearing areas and transfer shear from one side of the vertical joint to the other, even though there might not be a bond on the vertical interfaces. Horizontal reinforcement should be placed through the vertical faces to discourage vertical shrinkage cracks from forming in the immediate vicinity of the interface. It is desirable to construct the horizontal surfaces of the steps on a 1 1/2: 12 slope. This permits better consolidation of the concrete and allows the escape of air when the concrete under the horizontal step is placed last.

8
Bent Caps

8.1 GENERAL

For aesthetic purposes it is preferable to keep the bent cap within the limits of the box girder superstructure so that the soffit of the entire bridge is one large plain surface interrupted only by the columns. It is also aesthetically desirable to use as few columns as possible. These two conditions combined can sometimes require that the bent cap become a major structural member of quite large proportions. Although the depth of the cap may be limited, the width can be made as wide as necessary to satisfy design requirements. Ten-ft (3-m) cap widths are not unusual, and some have been much wider. Columns can be spaced at 30- to 40-ft (9- to 12-m) centers without undue difficulty, and considerably greater spacings can be obtained by prestressing the cap.

Limiting the depth of the cap to the depth of the superstructure has an advantage in single column bents that would otherwise be skewed. A hidden cap can be squared with the superstructure regardless of clearance requirements under the structure. A dropped cap with a single column bent could require a skewed superstructure in order to obtain vertical clearances under the structure. This would not be desirable from a design point of view; it would be more costly to construct and less pleasing aesthetically.

33

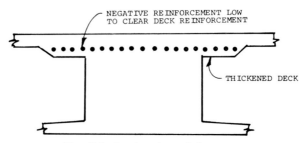

Fig. 8.1—Section through bent cap.

8.2 DESIGN ASSUMPTIONS

Bent caps can be designed as T-beams, assuming the deck slab acts as a flange for positive moments and the soffit slab a flange for negative moments. The effective width of the slab, according to AASHTO design specifications, is six times the slab thickness on each side of the cap. It is sometimes desirable, or necessary, to thicken the deck slab at the bents to provide space for locating the negative cap reinforcement beyond the ordinary limits of the cap (Figure 8.1). Thickening the deck and spreading out the cap reinforcement in this manner reduces the congestion of the reinforcing steel, gives the reinforcement a greater effective design depth, and also increases the area of concrete in compression.

The soffit slab is frequently flared near the bent in order to keep the main box girder compressive stresses to a desirable level for negative girder moments. This soffit flare is also helpful in resisting negative cap moments. Consideration should be given to flaring the soffit at the bent for designing the cap even if it is not required for resisting main girder moments.

It has been generally recognized that the design of bent caps, especially with large columns, has been unduly conservative. Recent investigations indicate that as much as 40 percent of the flexural reinforcement required by the older methods of designing caps can be eliminated, and the unit will still have an adequate factor of safety. Most of this reduction results from a change from Working Stress to the Load Factor method of design.[9] The following recommendations were also concluded from these investigations:

A. The effective width of an overhanging compression flange on either side of the web of an integral bent cap shall not exceed the following:

(1) One-tenth the span length of the bent cap. For cantilevers, the span used shall be twice the length of the overhang.

    (2) One-half the clear distance to the next bent cap.

    (3) Six times the least thickness of the slab.

B. The effective width of an overhanging tension flange on either side of the web of an intregal bent cap shall not exceed the following:

    (1) The effective width defined for compression.

    (2) One-fourth the average spacing of the intersecting box girder webs.

The effectiveness of reinforcement was found to decrease rapidly with distance from the bent cap web. This effect was assumed to be primarily due to shear lag. Reinforcement placed within the effective tension flange width can be assumed to be fully effective.

8.3 REINFORCING STEEL DETAILS

Special attention should be given to the details to insure that the column and cap reinforcement will not interfere with each other. This can be a problem especially when round columns with a great number of vertical bars must be meshed with a considerable amount of positive cap reinforcement passing over the columns. If the bridge is tangent or slightly skewed and the deck reinforcement is parallel to the bents, the negative cap reinforcement can be placed directly under aud in contact with the negative girder reinforcement (Figure 8.2). The deck distribution steel and "giveaway" bars can be stopped a few inches away from the cap reinforcing steel.

When the structure is on a greater skew and the deck steel is

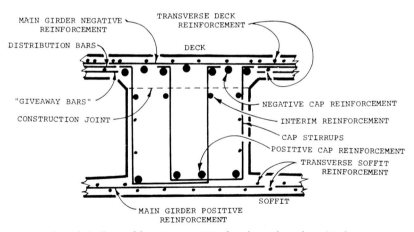

Fig. 8.2—Typical bent cap section for skews less than 20 deg.

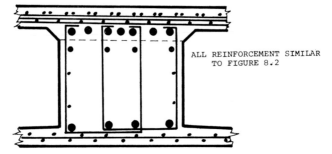

ALL REINFORCEMENT SIMILAR
TO FIGURE 8.2

Fig. 8.3—Typical bent cap section for skews more than 20 deg.

placed normal or radially to the longitudinal centerline of the bridge, the negative cap reinforcement should be lowered to below the main deck reinforcement or "giveaway" bars. This eliminates the necessity of having to custom cut all of the transverse deck reinforcing steel that crosses the bent cap (Figure 8.3).

8.4 SEISMIC REQUIREMENTS

For seismic conditions, vertical column reinforcement should be extended into the bent cap to develop the yield strength of the steel. If the superstructure is too shallow to accomodate the minimum anchorage length of a straight bar, the tops of the column bars can be hooked. It is also desirable that splices of main column reinforcement be avoided whenever possible. When the tops of the column bars are hooked, the details should provide for proper embedment of the tails in concrete. In some instances it may be desirable to increase the the width of the cap in the vicinity of the column in order to give the hooks proper coverage.

In single column bent caps, and at cantilevered ends of multiple column bents, it is advisable to place reinforcing steel at the bottom of the cap parallel to the cap even though none is required for design purposes. In the past it was customary to place #4 or #5 bars in the bottom corners of the cap stirrups. This nominal reinforcement has proven to be inadequate for earthquake conditions because it is not sufficiently effective in holding the concrete together to insure a bond between the cap concrete and the column reinforcement. Although no design criteria are available for determining the amount of reinforcement required, placing a #11 bar in each bottom corner of each stirrup is considered to be a minimum for an average structure.

8.5 INTERIM REINFORCEMENT

A horizontal construction joint is required near the top of the bent cap so that the deck slab can be placed at a later date. Either all or most of the negative cap reinforcement is usually located above the construction joint. After the cap concrete has set and before the deck concrete has hardened the falsework is subjected to compression, joint take-up, and settlement. This is similar to the situation discussed in Section 10.3. Although computing the amount of reinforcement required is very indeterminate, one rule of thumb, which the California Division of Highways has found to work satisfactorily, is to provide enough to support the weight of the cap plus the weight of 10 ft (3 m) of superstructure on each side of the cap—excluding weight of the deck. Sixty-seven percent overstress is permitted in tension and bond. Concrete compression and shear stresses are ignored for this condition only.

9
Hinges

9.1 GENERAL

In long continuous structures it is often not feasible to provide for all of the longitudinal temperature and shrinkage movement at the abutments. Unless the columns are unusually high or flexible, these movements produce high moments in the columns and cause serious design problems. In very long structures hinges are required every few hundred feet if the columns are relatively short. The distance between hinges can be increased if the columns are more flexible and capable of accommodating greater movements or if the superstructure is permitted to move on the tops of some of the columns.

As a general rule, expansion joints are the only portion of concrete box girder structures that require maintenance. It is virtually impossible to construct any type of expansion joint that does not produce a noticeable bump or sound in a vehicle that passes over it. It is also exceedingly difficult to construct a hinge that is completely watertight. Any type of hinge also increases the overall cost of a structure. For these reasons it is desirable to keep the number of hinges and joints in a structure to an absolute minimum.

From a design standpoint, hinges should be located near the point of inflection. For continuous box girder bridges with approxi-

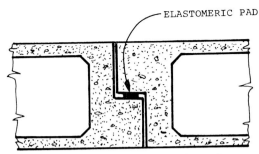

ELASTOMERIC PAD

Fig. 9.1—Typical elastomeric pad-type hinge.

mately equal spans, the most desirable location is usually at about the 1/5 point of the span.

Hinges are one of the weakest links of a concrete box girder bridge subjectd to earthquake forces. They should be designed to accommodate daily and seasonal changes in length of a structure, but if they allow too much movement during an earthquake, the entire structure can be destroyed. Since the hinge is such a vital detail it should be designed conservatively. No two earthquakes are alike, and seismic design criteria are only approximations. Columns can deflect considerably more than calculated for design forces without failing completely. Ground movements can cause unrestrained hinges to open up without bending the adjacent columns to the point of failure. Hinges should be detailed so that they cannot come apart and allow the suspended portion to drop.

One of the most commonly used types of box girder hinges is similar to Figure 9.1, except that two steel angles, placed back to back, are used for the horizontal sliding surface. A thin sheet of asbestos, lead, or other suitable material is usually used to lubricate the surface and facilitate sliding. There is a tendency for these joints to "freeze" because of grout leaks or other construction irregularities. This type of hinge is considered to be especially inadequate if earthquakes are any consideration at all. Considering variations during construction, shrinkage, extreme temperature shortening, movements due to adjacent hinges freezing, and other unpredictable movements, even the largest size of commercially available angles is not considered to have an adequate bearing length for any box girder bridge bearing. Steel plates have been used in lieu of steel angles in order to provide a longer sliding surface, but they still retain all the other disadvantages. Since the plates are rigidly embedded in the concrete, they rotate as the spans deflect, and high edge loadings result.

The elastomeric pad-type hinge, illustrated in Figure 9.1, offers very little resistance to movement and also accommodates differential rotation between the two sides of the hinge. Although this hinge has a tendency to "freeze" in some installations, it has quite a good service record. In spite of its good performance, it has been criticized because it is practically impossible to inspect or replace the elastomeric pads.

Unless positive means are taken to prevent it, dirt and water from the bridge deck fall into this hinge. The dirt is compacted by successive movements of the hinge so that it does not work as designed. Water stains the sides and undersurface of the bridge superstructure. A number of deck joint devices are available that keep dirt out of the joints, but not any are known at the present time that are entirely effective in keeping out water for any length of time without constant maintenance.

When the underside of a hinge is constructed after the overhanging portion is completed, it is difficult to remove all the air and properly consolidate the concrete under the bearing pads. Consistently good results have been obtained by requiring that the underside always be completed first. If the contractor's operations favor constructing the suspended side of the span first, a construction joint can be made several feet from the hinge, and that short section, containing the upper portion of the hinge, can be placed after the cantilever side is completed.

Figure 9.2 illustrates an improvement of the hinge shown in Figure 9.1. Concrete bolsters are cast with the diaphragms on each side of the hinge. Elastomeric pads, rockers, or other suitable types of bearings can be used to transfer the vertical loads and accommodate the longitudinal movement between the two sides of the hinge. The bolsters can be in line with the girder stems in reinforced

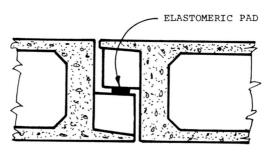

Fig. 9.2—Elastomeric pad-type hinge that is an improvement over the type shown in Fig. 9.1.

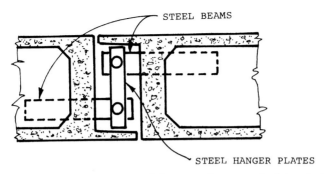

Fig. 9.3—A hinge employing a steel hanger plate attached to steel beams cast in the adjoining girders.

concrete box girder structures, or, if the girders are prestressed, the bolsters can be offset from the girder stems so that they do not interfere with the prestressing anchorages. Access can be provided, if desired, so that the bearings can be inspected and repaired if necessary.

In Figure 9.3 vertical loads are transferred from one side of the hinge to the other by means of steel hanger plates attached to steel beams cast in the ends of the adjoining girders. Access for inspection can be readily provided by means of openings in the cantilever soffit slab and possible additional openings through the hinge diaphragms and girder stems.

The effectiveness of the hinges in Figures 9.2 and 9.3 are little affected by water and dirt falling through the opening in the deck. It is recommended that this be permitted for the sake of economy and maintenance unless water is not allowed to fall on traffic or facilities underneath the bridge. Waterstops or other suitable types of joint seals can be used if required.

The three hinges illustrated are only a few of the numerous types that have been used in connection with box girder bridges.

A number of newly or partially completed box girder bridges were near the epicenter of the February 9, 1971, San Fernando earthquake. Some of these bridges were completely destroyed or damaged to varying degrees. The hinges in the superstructures were one of the most significant factors contributing to the failure of the structures. Without a significant amount of longitudinal restraint through the hinges, adjacent frames moved independently and in different directions. When the relative movement between adjacent frames equalled the width of the 14-in.-long (356-mm) hinge seats, the suspended portions of the spans lost their vertical support and dropped.

The surface of the earth during an earthquake acts much like ocean waves. Since it is not practical to tie all bridge footings together to make the entire structure act as a rigid unit, the distance between footings may vary as the earth's surface undulates. Each footing can also move in any direction or rotate about any axis independently of the movements of any other footing.

A field study was conducted after the 1971 San Fernando earthquake in an area surrounding some of the bridges that were severely damaged. Many survey points had been placed on known coordinates with a high degree of accuracy to establish property lines, easements, and alignment controls before the earthquake. The area selected for this study excluded regions where faulting or earth slides were evident. The directions of the lines were at random and did not necessarily coincide with the direction of maximum distortion of the earth. Ten different lines with an average length of 136 ft (41 m) either increased or decreased in length an average of 1/2 percent. The actual changes in lengths of these lines during the earthquake were undoubtedly much greater. Vertical displacements were not measured. These values cannot be used quantitatively for design purposes, but they can give a feeling for what is required when designing hinges and restraining devices.

Hinges are one detail that probably provide more opportunity for improvement than any other part of a box girder bridge. It is considered good practice to design a structure so that its strength is governed by the strength of the main members and not by the strength of the connections or details. It is recommended that the calculated loads on a hinge be increased by a factor of 4/3 for design purposes in order to assure that it will not fail before the main members.[10] Also, the design should include a horizontal longitudinal force on the hinge in excess of what is theoretically expected of the bearing device.

9.2 EQUALIZING HINGE MOVEMENTS

In long continuous box girder viaducts and river bridges that have a number of hinges, it is quite common for some hinges to take very little or no movement at all while other hinges move much more than planned. This is due to grout leaks, binding surfaces, or other construction defects. This has been experienced even in structures that have hinge bearings with very low coefficients of friction or low shear displacement values. If one hinge does not function, the movement it is intended to accommodate must be taken by adjacent hinges. This results in some columns being deflected more than the designer intended, and the entire frame analysis is erroneous.

"Equalizing bolts" have been successfully used to insure that each hinge takes its fair share of the total longitudinal movement. Equalizing bolts are basically large steel bolts with washers placed through the hinge and adjusted so that they are loose until the hinge opens up to its designed maximum width. The bolts then come into action and prevent the hinge from opening farther. Any additional movement in the structure must then be accommodated by the next adjacent hinge.

Since the forces in hinges that resist normal movement are due to accidental causes, it is not possible to design the equalizing bolts for known forces. A 1 1/2-in. (38-mm) round ASTM A–325 steel bolt placed through the hinge in each cell has proven satisfactory. Some difficulties have been experienced because the nuts were brought up tight before the hinge had a chance to accommodate even the initial shortening due to temperature and shrinkage. This occurred because the last workmen who had access to the cells felt that the nuts should be tightened. In order to eliminate this problem it is now customary to specify soft washers with a thickness equal to the desired gap. Expanded polystyrene, which has a crushing strength of approximately 35 psi (2.5 kgf/cm²), has been used satisfactorily.

It is desirable to have the equalizing bolts start acting as soon as a hinge opens to its maximum design width and to develop their full strength without stretching appreciably. This is in contrast to the function of hinge restrainers for resisting earthquake forces (discussed in Section 9.3), which must stretch in order to absorb energy For this reason it is recommended that equalizing bolts or similar devices be used in addition to hinge restrainers when a structure contains two or more hinges.

9.3 RESTRAINERS—NEW STRUCTURES

If the various units of a long bridge, which are separated by hinges, would sway in phase with each other and have the same displacement, during an earthquake, there would be no problem if all other factors were favorable. This condition is obviously not practical. Restraining devices are required to limit the amount that a hinged joint can open. It is also desirable for the restraining devices to absorb some of the energy that the earthquake imparts to the structure.

An ideal hinge restrainer would permit the hinge to accomodate the normal daily and seasonal changes in length without offering any appreciable resistance, but resist any sudden changes. It should

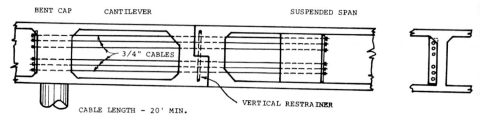

BENT CAP   CANTILEVER   SUSPENDED SPAN

3/4" CABLES

CABLE LENGTH - 20' MIN.   VERTICAL RESTRAINER

Fig. 9.4—Earthquake hinge restrainer.

preferably be able to stand by for years without any maintenance and be fully capable of doing its job when required. Such shock absorbers can be made but are relatively expensive. More economical devices can be devised for the more ordinary structures, but shock absorbers should be considered for unusual bridges.

Since the job of a restrainer is to keep sections of a bridge from moving in different directions or from moving in the same direction with different velocities at any given time, it is necessary for the restrainer to expend energy. Energy is the product of force multiplied by displacement, or the area under the load-elongation curve. If a restrainer is capable of resisting the required force but is not capable of stretching the distance necessary to absorb the required energy, it will obviously fail. It is therefore essential that a hinge restrainer be constructed of ductile materials. A restrainer's ability to just resist a great force is not good enough.

Groups of straight steel cables are currently being used in California for tying segments of structures together at the hinges (Figure 9.4). The cables are adjusted so that they start to act as soon as the hinge reaches its maximum normal open position. They are a minimum of 20 ft (6 m) long in order to absorb the energy required to redirect adjacent segments as they move away from each other. A sufficient number of cables must also be used to limit the maximum opening of the hinge. Although these cable restrainers have not yet been tested by actual earthquakes, they have been subjected to dynamic analysis and model testing and appear to be satisfactory. Dynamic analyses and models have shown that segments of structures without these restrainers would "fly apart" during an earthquake. It is likely that some spalling may occur as a hinge closes abruptly, but the damage should not be serious or costly to repair. In some cases it may be desirable to use elastomeric pads or similar type buffers to reduce the force of the impact. On structures where repairs or replacement would cause extreme inconvenience, more elaborate and expensive devices would be justified.

When designing a hinge restrainer unit to resist seismic forces, it is important to remember that the device must absorb or dissipate energy as well as provide a force to hold the units of the structure together. The California Division of Highways assumes a restrainer should resist a force equal to at least 25 percent of the weight of the smaller segment of structure involved, and the elongation should not exceed 2 in. (50 mm) at working stress level. Dynamic analyses for some bridges in specific locations have indicated that this 25 percent minimum force may be very inadequate.

In plan view, a long structure acts as a horizontal beam. Restrainers should be concentrated as much as is practical at the outer edges of the hinge in order to give the structure maximum resistance to transverse bending.

9.4 VERTICAL RESTRAINERS

Box girder bridges at the site of the San Fernando earthquake, February 9, 1971, showed evidence of vertical forces that contributed to the damage of the structures. The California Division of Highways now ties the upper and lower portions of box girder hinges together with 3/4-in. (19-mm) cables (Figure 9.4). These cables permit longitudinal movement of the hinge but restrain vertical separation of the joint.

9.5 TRANSVERSE SHEAR KEYS

Shear keys to prevent transverse differential movements in box girder hinges can be made with very little, if any, additional cost. They are essential for earthquake forces and desirable for keeping the bridge in alignment and resisting differential movements due to wind, traffic, temperature, and other forces. Unless better design information is available, it is suggested that transverse shear keys be designed for a minimum force equal to 25 percent of the contributing dead load. It should be remembered that the ductility factor of an ordinary concrete shear key is close to unity.

9.6 RESTRAINERS–EXISTING STRUCTURES

Many completed structures with hinges, which are considered to be in critical locations, have been retrofitted with hinge restrainers. The design criteria are the same as used for new construction. Access to both sides of the hinge is made by cutting holes in either the deck or soffit slab, depending on structural problems and accessibility.

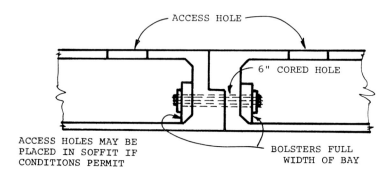

ACCESS HOLES MAY BE
PLACED IN SOFFIT IF
CONDITIONS PERMIT

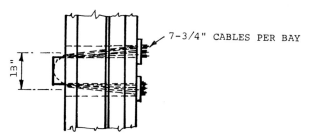

Fig. 9.5.—Earthquake hinge restrainer for existing bridge.

Holes are then cored through the hinge and the two sides of the hinge tied together by the use of steel cables and anchor plates (Figure 9.5).

One of the critical design considerations is to insure that the cables will not tear out the hinge diaphragms. This occasionally requires additional restrainers with fewer cables per restrainer unit or strengthening the diaphragms with reinforced concrete bolsters.

10
Girder Reinforcement

10.1 SELECTION AND ARRANGEMENT

In structures designed by the Working Stress Design method, #11 bars are the most commonly used size of girder reinforcing steel. Occasionally, #14 and #18 bars are used when the number of smaller bars causes undesirable congestion. Structures designed by the Load Factor Design method require less total girder reinforcement and favor the use of smaller size bars—usually #9 and #10.

There is a tendency for cracks to form in concrete at the ends of tensile reinforcing steel. If a number of bars are stopped at the same location, the cracks tend to join each other and produce a larger crack. For this reason it is considered good practice to stagger the ends of reinforcing bars as illustrated in Figure 10.1.

It is considered good practice to use as few bar lengths as possible in order to simplify shop detailing, cutting, handling, and placing. Maximum bar lengths should be kept shorter than 60 ft (18 m) whenever possible in order to facilitate handling. Sixty-ft (18 m) lengths are commonly available in most areas. It is usually possible to use many bars of the same length by adding a foot or two to a few of the bars. The additional cost for the slight amount of extra steel is offset by the savings in fabricating, handling, and placing the number of different lengths. This condition is usually obtained by

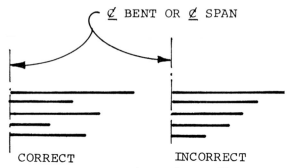

Fig. 10.1—Reinforcing bar cutoffs.

combining the longest cutoff on one side with the shortest on the opposite side. A few odd lengths are usually required. Applying this scheme to the negative bar cutoff lengths (in feet) on the moment diagram in Figure 10.2:

37	29	23	19	16	13	10	8
8	12	10	14	17	19	23	25
45	41	33	33	33	32	33	33
					(use 33)		

Thus by adding 1 ft to one of the bars, six identical length bars are obtained.

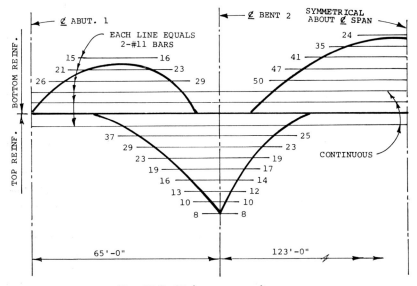

Fig. 10.2—Girder moment diagram.

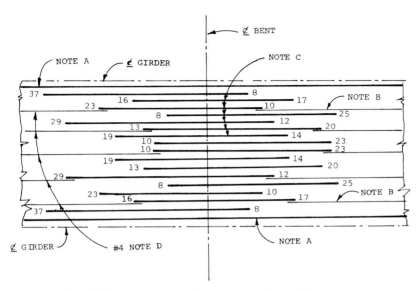

Fig. 10.3—Arrangement of negative girder reinforcement.

The layout for the negative girder reinforcement in one bay of the structure is shown in Figure 10.3. Detailing of the positive reinforcement is similar.

In reinforced box girder structures it is customary to place two continuous girder bars at the bottom of the stem stirrups. However, when 8-in. (200-mm) stems are used and the transverse soffit reinforcement is hooked at both ends, the out-to-out length (overall dimension in one direction) of the hooked bars may be somewhat less than the theoretical, and the hooks decrease the width available inside the stirrups for the two continuous bars. To overcome this problem, it is common practice to place only one of the two continuous bars inside the exterior girder stirrups, with the other bar outside but next to the stirrups.

Although the design specifications that limit the effective width of flanges apply only to compression flanges, some designers have felt it should apply to main girder reinforcement in the tension flanges also because of the shear lag effect. Accordingly, many box girder bridges have been constructed with all the main girder reinforcement placed within a distance of six times the slab thickness from the face of the girder stem. As the trend for longer spans and thinner structures grew, the girder reinforcement became unduly congested within these limits. With the experience gained from a great number of box girder bridges, designers confidently spread the girder reinforcement throughout the entire width of the soffit and deck slabs, and that has become an accepted practice. According to folded plate analysis, the shear lag effects are insignificant for ordinary highway interchange structures as long as practical depth-to-span and girder spacing-to-span ratios are used.

10.2 TRANSITION REINFORCEMENT

At one time it was thought that a smoother flow of stress between reinforcing steel and concrete could be obtained by splicing #8 or #9 reinforcing bars to the ends of #11 or larger main girder reinforcement. These "transition" bars were used in many structures over a period of years. A careful examination of these structures compared with structures built without the transition bars led to the conclusion that there were no visible benefits obtained by splicing smaller bars to the #11 or larger bars.

10.3 INTERIM REINFORCEMENT

The deck slabs of box girder bridges are customarily constructed some time after the soffit slab and stems have been completed. During the interval of time after the stem concrete has taken its initial set and the girder reinforcement in the deck is capable of resisting longitudinal stresses, any settlement of the falsework, compression of the falsework members, or additional take-up in the falsework joints produce negative moments in the uncompleted superstructure consisting of the soffit slab and stems. These moments can produce tensile stresses and cracking in the girder stems unless reinforcement is placed immediately below the horizontal construction joints at the tops of the stems. Since these settlements are dependent on the foundation conditions, type of falsework, quality of workmanship, and various other factors, it is not possible for the designer to design the required reinforcement with the same degree of accuracy as the main reinforcement. AASHTO specifications require that interim reinforcement consist of 10 percent of the negative reinforcing steel, or two #8 bars placed at the top of the stem for the full length of the girders. Two #8 bars have proven to be very satisfactory for a great variety of structures over a considerable period of time. It has been found necessary to locate one of the bars approximately 4 in. (100 mm) below the other, at the tops of 8-in.-wide (200-mm) stems, in order to permit a vibrator to function properly.

11
Load Distribution

11.1 BACKGROUND

Early box girder bridges in the United States were designed as modified T-beam structures, and the same formulas were used to assign live loads to box girders. As designers gained confidence in the reliability of box girder construction they reasoned that the addition of a soffit slab to a T-beam aided in distributing live loads to the girders. This led to a revision of the specifications, which permitted the live load on a box girder bridge to be spread out to a greater number of girders than would be permitted on a T-beam bridge with a similar girder spacing.

Many box girder bridges have been built with very little deck overhang beyond the exterior girders. According to older AASHO specifications, the design shear capacities for those exterior girders were considerably less than for interior girders. Many of those exterior girders developed visible diagonal tension cracks near the supports, which led to the conclusion that exterior girder stems should have at least the same shear capacity as interior girders.

11.2 UNIT DESIGN

With the advent of posttensioned box girder bridges and the use of large tendons, the California Division of Highways adopted the practice of designing the entire superstructure as a unit rather than as a number of individual girders. This was economically and practically desirable in order to make full use of large tendons. With these assumptions, alternate girders can have a different number of tendons tensioned to full capacity. This design practice was extended to reinforced concrete box girder bridges. When reinforced box girder bridges were designed as individual girders with a considerable deck overhang, the positive girder reinforcement was unduly congested in the exterior cells. The unit design method permits distributing all the girder reinforcement uniformly throughout the width of the structure. Also, all girders are designed with the same shear capacity.

11.3 GIRDER PARTICIPATION

Both experimental and mathematical analyses indicate that girder webs close to a column transmit substantially more shear to the bent cap than webs of girders that are farther from a column. This is because bent caps deflect when subjected to dead and live loads, and girders near the columns are more rigidly supported. At loads substantially greater than the service load, in particular near ultimate, experimental data indicate that the shears transmitted from the box girder webs to the bent cap are approximately uniform.[9]

For design purposes it is assumed that bent caps do not deflect and all girders carry the same loads regardless of their relative positions on the bent caps.

12
Dead Load Deflection
and Camber

12.1 DEAD LOAD DEFLECTION

It is important to be able to predict the deflection of a concrete bridge span due to its own weight. The predicted dead load deflection is one of the factors used to determine the amount of camber to be placed in the falsework.

Dead load deflections are affected by the age and strength of concrete; the amount and type of stress in the concrete; when the load is applied; the amount and placement of reinforcing steel; the type, quality, and proportions of materials in the concrete; the method of curing; and other factors. Most of these factors can be controlled so that deflections can be predicted with reasonable accuracy.

The deflection of a concrete span increases with time. The percentage of the ultimate deflection at various ages is approximately:

25 percent when falsework released—continuous spans
33 percent when falsework released—simple spans
70 percent at 6 months
82 percent at 1 year
90 percent at 2 years
93 percent at 3 years

53

95 percent at 4 years
100 percent at ultimate

It is customary to require that concrete attain a specified strength before supporting falsework is removed. If all other conditions are the same, a concrete span that has its falsework removed at an early date will have greater initial and ultimate deflections than a similar span supported by falsework for a longer time. It is necessary to make compromises in order to keep deflections within reasonable limits and to construct a bridge within a reasonable length of time. For ordinary cast-in-place designs and construction conditions it is suggested that falsework be kept in place for at least 10 days and the concrete attain at least 80 percent of its designed ultimate strength before the false-work is released. The falsework should be released and the structure allowed to support its own weight as soon as both of these conditions are satisfied. If it is planned to have the falsework support the structure for a considerably longer period of time, less camber should be provided.

Dead load deflection of segmental prestressed bridges is discussed in Section 20.5.

12.2 CAMBER

Camber is the correction made to the vertical alignment of the forms to compensate for the anticipated movements that take place when the concrete is placed, during the curing period, and after the falsework is removed.

If the falsework and forms for a concrete bridge were built according to the final desired grade of a structure without making any corrections, the resulting structure would in all probability have an undesirable profile grade, unsightly sagging spans, and poor deck drainage.

Since a reinforced concrete span continues to deflect over a long period, it is necessary to decide whether the initial camber should be selected on the basis of whether the optimum profile grade should be realized as soon as the bridge is completed or at some later date. Under ordinary circumstances, if the depth-to-span ratio of the structure is not excessively low, satisfactory results can be obtained by planning for the ultimate conditions.

Forms and falsework tend to settle and compress under the weight of freshly placed concrete. Joints in the falsework take-up and the falsework footings may compress the underlying soil. The amount of this movement is dependent on the type and design of the falsework, workmanship, type and quality of materials, and footing conditions. When the falsework is removed, the spans will generally sag from the dead load, but in the cases of extremely unbalanced spans or short cantilevered sections, upward movements might be experienced.

When the alignment of corners and surfaces of concrete decks and bridge rails are not exactly true, small variations are accentuated in foreshortened views and detract from the overall appearance of the structure. In order to minimize these irregularities it is advisable that all work above the level of the bridge deck be completed after the falsework has been removed.

13
Live Load Deflection and Vibration

Live load deflections and vibrations are often interpreted by the public to be an indication of structural weakness. Complaints from individuals concerning observed deflections and vibrations usually express the opinion that the structure is not safe.

All structures deflect when subjected to live loads. Live load deflections of almost any bridge can be detected by standing at mid-span as a heavily loaded truck or train crosses the bridge. Concrete box girder bridges, however, are comparatively rigid and deflect less than other types of construction with similar spans and structure depths.

Moving pedestrians can cause a bridge to vibrate, but the movements of a highway bridge due to pedestrians is generally imperceptible. Marching in cadence, however, can cause concern, and marching troops have historically broken step when crossing bridges.

Individuals vary in their reactions to vibration. The two main factors, amplitude and period, combined with the varying sensibilities of individuals make it impossible to establish exact limits for perceptibility.

Vibration of concrete box girder bridges is not a structural problem. No known reported failures or problems with concrete box girder bridges are attributable to vibrations. Vibration problems of bridges are primarily psychological. Humans tend to exaggerate any movement or vibration, and engineers who have investigated these responses estimate that the amplification factor can be in the order of 100 to 1.[11]

It is not likely that occupants of vehicles can detect the vibrations of a bridge unless the vehicle is parked, in which case the suspension of the vehicle usually reduces the sensation. However, if the periods of vibration of the vehicle and bridge span are the same, amplification is possible. Pedestrian insecurity is the determining factor. Most complaints concerning vibrations are on structures other than concrete box girders.

The live load deflection of a span, by itself, cannot be used to determine whether or not vibrations might be objectionable. Considering the passage of a single loaded vehicle across a bridge, the span has zero deflection as the vehicle approaches the bridge, attains maximum deflection when the vehicle is near the center of the span, and returns to zero as it leaves the span. Considering the velocity of the vehicle, the frequency of this deflection will be well below one cycle per second except for very short spans. Human sensitivity to such low frequencies is very slight.

The magnitude of stresses in concrete box girder bridges due to vibration is very low. The stress range in reinforcing steel is so low that there should be no danger of fatigue due to cyclic loading.

Dynamic analyses of box girder bridges have indicated that very long span bridges have vertical periods of vibration that can become resonant with the vertical movements of earthquakes. Analyses of some reinforced concrete spans have indicated that high tension stresses may occur in the nominal bottom girder reinforcement at intermediate supports and top girder reinforcement at midspan. Although the preliminary information is too limited to draw positive conclusions, it appears that it might be given consideration in long span prestressed structures.

14
Thermal Effects

14.1 GENERAL

AASHTO bridge design specifications give factors to be used for determining changes in lengths of members due to temperature changes. It has generally been assumed that all parts of a structure are the same temperature and that there are no appreciable thermal variations within the structure at any given time; the only effects of temperature relate to expansion and contraction of the entire structure, which cause flexure of the columns and working of the expansion joints.

Actually, all portions of a bridge are not the same temperature at any given time. Variations in temperature result in dimensional changes that produce moments and shears throughout the entire structure. Although stresses due to these thermal variations have generally been ignored, they have, in some instances, caused serious cracking and opening of joints on foreign bridges.[4, 12, 13, 14]

Temperature variations within a bridge are caused by exposure of the deck to the sun, the temperature of the air within the girder cells, and the temperature of the surrounding atmosphere. These factors act out of phase with each other.

A limited amount of work has been done with placing heating elements in or on bridge decks and heating the air inside of box girder cells in order to prevent icing of the decks. This type of heating also causes temperature variations within the structure but so far has not caused any difficulties.

Blacktop surfacing on a bridge deck absorbs more heat than a light-colored concrete surface. Temperature differentials as high as 55 F (33 C) have been measured between a deck with blacktop surfacing and the soffit slab. A light-colored deck can reduce this differential by as much as 15 F (8 C).[15, 16]

14.2 LONGITUDINAL

The overall longitudinal changes in length due to temperature variations are less for concrete box girder bridges than for other types of concrete bridges. This is owing to the insulating features of the trapped air volume inside the box structures.[17]

It is obvious that the sun shining on a bridge deck causes the deck to become considerably warmer than the soffit. As a deck becomes heated and expands faster than the soffit, the span has a tendency to arch upward. This arching action of the spans produces positive moments throughout the spans of continuous structures (Figure 14.1). These moments have ordinarily been considered negligible and consequently ignored.

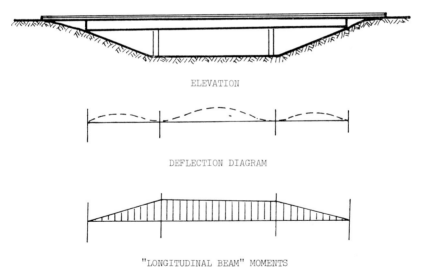

ELEVATION

DEFLECTION DIAGRAM

"LONGITUDINAL BEAM" MOMENTS

Fig. 14.1—Elevation, deflection diagram, and "longitudinal beam" moments. Effect of a warm deck and cool soffit on a continuous structure.

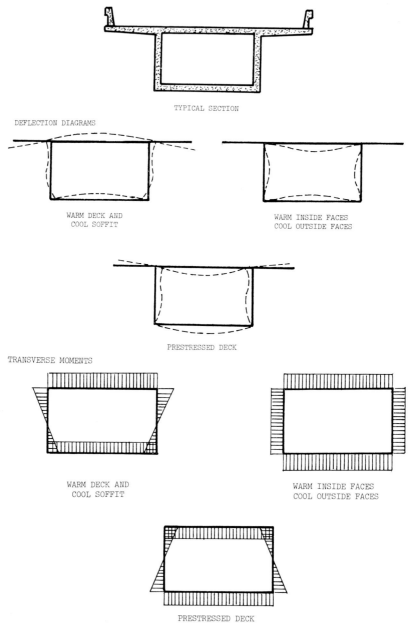

TYPICAL SECTION

DEFLECTION DIAGRAMS

WARM DECK AND
COOL SOFFIT

WARM INSIDE FACES
COOL OUTSIDE FACES

PRESTRESSED DECK

TRANSVERSE MOMENTS

WARM DECK AND
COOL SOFFIT

WARM INSIDE FACES
COOL OUTSIDE FACES

PRESTRESSED DECK

Fig. 14.2—Deflection and transverse moments developed in a typical section due to temperature differences between air inside and outside of box girder.

In continuous prestressed bridges the prestressing forces produce positive moments at the supports and negative moments at midspans. If the concrete stresses in the soffit are nearly zero at the intermediate supports, for dead load plus prestress, and there is moreover an insufficient amount of reinforcing steel, a warm deck and cool soffit can cause tensile stresses in the soffit near the supports that result in transverse vertical cracks. This can be particularly critical in segmental type construction if there is very little or no reinforcing steel passing through the joints between the segments.

Cracking of the soffit near intermediate supports due to temperature can also be aggravated by other factors. If the points of contraflexure of the prestressing tendons are farther than necessary from the bents, the length of soffit over which cracking occurs is extended. Also, highly concentrated bearing forces produce longitudinal tensile stresses in the soffit near the support.[4, 16]

Almost all of California's prestressed box girder bridges are of the cast-in-place type construction, and the span lengths range up to 300 ft (90 m). The soffit slabs are longitudinally reinforced with a minimum of 0.3 percent of conventional reinforcing steel. Superstructure reactions are usually transmitted to the piers through large diaphragms; this avoids highly concentrated loads under the girder stems. No cracks have been observed in any of these structures that can be attributed to temperature differentials.

Some bridges in other countries have had vertical transverse cracking extending from near the intermediate supports to the point of zero moment. These bridges do not have conventional reinforcement in the soffits. The cracks extend to midheight in box girder structures and nearly to the deck slab in T-beams. The formation of the cracks is attributed to "longitudinal beam moments" due to temperature differentials aggravated by longitudinal stresses caused by high bearing reactions. The concrete at these locations had zero or even slight tensile stresses under dead load plus prestress and uniform temperature conditions.[16]

Studies have shown that a temperature difference of 18 F (10 C) between the deck and soffit slabs can be a considerable tension-producing factor in some prestressed box girder structures.[18] Temperature differentials can in reality be considerably greater.[12, 13, 19]

14.3 TRANSVERSE

In addition to the "longitudinal beam" stresses, box girder bridges are also subject to transverse bending moments caused by temperature differentials between the deck and soffit slab.

Stresses due to temperature differentials between the inside surfaces of the girder cell and the outside surfaces are not ordinarily calculated. However, serious longitudinal cracking has been experienced in the bottoms of the girder stems in a number of foreign cantilever box girder bridges. Field measurements have shown that great temperature variations do exist throughout the concrete sections, and mathematical analysis confirms the suspicions that these temperature differentials are responsible for the failures.[16, 20]

As the sun shines on a box girder structure during the day, the air inside of the cells and the inner faces of the cells can become quite warm. If the outside air then becomes quite cold, there can be a considerable temperature variation in the deck, walls, and soffit which produces tension stresses in the outer faces and compression in the inner faces (Figure 14.2). If members of minimum thickness are used at midspan, they will probably be flexible enough to resist the minor stresses without cracking. At the supports, however, where the girders and soffit are flared and are quite stiff, the stresses can be of a considerable magnitude. This condition is aggravated since there is also a considerable amount of horizontal shear at this location. Also, if the deck slab is prestressed transversely, additional tensile stresses are created in the outer faces of exterior girders near the soffit, which makes the situation even worse.

These problems can be minimized by keeping the thickness of the members to a minimum, reinforcing the girder stems with more stirrups in these critical areas, and placing vents in the soffit and near the tops of the girder stems in order to keep the temperature inside the cells as near to the temperature of the outside air as possible.[4, 16]

To date, there are no known difficulties of this type experienced with the cast-in-place box girder bridges so commonly used in the United States.

15
Reinforcing Steel

15.1 GRADES OF STEEL

Grades 40 and 60 reinforcing steel are both used in concrete box girder construction today. The permissible working stresses are 20,000 and 24,000 psi (1400 and 1690 kgf/cm^2), respectively, for WSD, with larger resulting stresses if the LFD method is used. Grade 60 reinforcing steel became popular in concrete box girder bridge construction in the 1960s and is now used exclusively in many areas.

Even though Grade 60 reinforcement has greater yield and ultimate strengths than Grade 40, the modulus of elasticity of the steel is the same, and increasing the working stresses also increases the total width of cracks in the concrete. In order to overcome this problem, LFD bridges generally have smaller, more closely spaced bars. Grade 60 reinforcement is usually not as ductile as Grade 40 and is more difficult to bend. This is particularly noticeable in field bending. Grade 40 steel often can (but should not) be bent to a smaller radius than required by the ASTM tests without any visible signs of cracking or breaking. Grade 60, however, is more susceptible to cracking or breaking if not bent with proper equipment or if bent to a radius smaller than recommended.

15.2 CONCRETE COVER

For normal exposures, 1-in. (25-mm) cover over reinforcing steel is satisfactory at the interior and exterior faces of box girder stems.

When girder stems are flared near the supports, in order to resist the greater vertical shear forces, coverages of up to 3 1/2 in. (89 mm) are permitted at the interior faces in order to avoid the necessity of fabricating, handling, and placing many different widths of stirrups. Cover over reinforcing steel in the exterior faces of girder stems should be increased to 3 in. (76 mm) if the bridge is located in a marine environment.

Reinforcing steel should have a minimum of 1 in. (25 mm) at the upper surface and 1 1/2 in. (38 mm) at the undersurface of the soffit slab for any environment.

Coverages for deck reinforcement are discussed in Section 3.3.

15.3 SPLICING

Reinforcing bars #11 and smaller are generally spliced by lapping. Under certain conditions they may be butt welded or joined by mechanical couplers. But #14 and larger bars should not be spliced by lapping. Butt welding reinforcing bars requires qualified welders and, under some conditions, may be relatively expensive. There are several types of mechanical couplers that give satisfactory results and are available commercially. The California Division of Highways requires that the total slip of the reinforcing bars within a mechanical splice sleeve, after loading in tension to 30,000 psi (2100 kgf/cm^2) and relaxing to 3000 psi (210 kgf/cm^2) shall not exceed 0.010 in. (0.25 mm) for #14 or smaller reinforcing bars, or 0.030 in. (0.76 mm) for #18 bars. The splice must also develop not less than 90 percent of the specified minimum ultimate tensile strength of the unspliced reinforcing bar. Splices of #14 and #18 bars are also staggered a minimum of 5 ft (1.5 m). Since mechanical couplers are thicker than the bar being spliced, the normal concrete cover is reduced when mechanical couplers are used.

In long spans where commercially available lengths of bars are not long enough and must be spliced, the splices should be located in areas where the laps or mechanical splice devices will not create undue congestion.

15.4 LARGE BARS

Large, #14 and #18, reinforcing bars have been used extensively for main girder, cap, and column reinforcement in box girder structures. They provide a large area of steel where a greater number of smaller bars would be unduly congested. Some reinforcing steel placers are opposed to the use of #14s and #18s because they are so heavy and require additional men or equipment for handling. A designer should carefully consider the alternatives before specifying these large sizes. This is especially true for small jobs or if very small quantities are required. In case of doubt, a local supplier should be consulted concerning the availability at a specific site.

15.5 COLUMN REINFORCEMENT

Column reinforcement, when extended into the footings and/or caps, should have adequate anchorage for developing the yield strength of the steel. The minimum embedment lengths for #14 and #18 column bars may be greater than the footing and cap thicknesses required by other considerations.

It is desirable that splices of main column reinforcement be avoided whenever possible. This is especially true for seismic considerations, and also if #14 and #18 bars are used. If column bars are hooked at both ends, the details should permit the hooks to be bent in the same plane in order to simplify fabrication and placement. If the details call for the hooks to be placed parallel to each other in the cap, in order to enclose the tails in cap concrete, and radially in the footing, the hooks for each bar must be bent at different angles in relation to each other. The problems involved in accomplishing this are expensive and not justified.

16
Prestressing

16.1 COMPARISON WITH REINFORCED BOX BRIDGE

From outward appearances it is difficult to determine whether a box girder bridge is reinforced or prestressed. The only difference, in most cases, is that the prestressed structure has a thinner super-structure than a reinforced bridge with the same span. Longer spans are more likely to be prestressed.

The details for reinforced and prestressed box girder bridges are very similar. The main difference is that the reinforced bridge has considerable longitudinal reinforcement in the deck and soffit slabs, whereas the prestressed bridge has large prestressing tendons draped in the girder stems. The same design is used for the decks of either type, and the economics of girder spacing and cantilevering of the deck beyond the exterior girders is the same. The girder stems of prestressed boxes must be thicker than the stems of reinforced boxes in order to accommodate the large prestressing ducts.

16.2 DETAILS

When bridge projects are constructed under the competitive bid-ding system, prestressed bridges are usually designed and detailed to accommodate any of the commonly available prestressing methods.

To take full advantage of the economy of large tendons it is desirable to design the entire superstructure as a unit and to permit, if necessary, a different number of tendons in alternate girder stems. For ordinary highway structures it has been found satisfactory to allow a variation of force between girders in the ratio of 3:2, if the force variation does not exceed 750 kips (340,000 kgf). The force is computed by assuming the stress in the strand to be $0.6f_s'$ and the tendons should be arranged symmetrically in the structure. Tendons should be stressed in a transverse sequence in order to minimize high concentrations of forces and transverse moments in the superstructure.

A considerable amount of difficulty has been experienced due to welding in the vicinity of prestressing steel. Spatter from welding or welding with the ground attached to a reinforcing bar or a prestressing tendon can pit the tendon by arcing at some distant point. Very minor pitting can cause failure of the tendons at a very low stress. Damage of this type can be done in a matter of minutes and may be difficult to detect unless there is a positive reason for looking for it. If this type of damage is suspected in a completed structure it may be virtually impossible to prove it or correct it. Details that require welding near a prestressed structure should be avoided, and welding equipment should not be permitted in the vicinity of prestressing materials or prestressed construction.

Posttensioning bridges from one end only should be given consideration for the sake of economy. It is generally advantageous to do so for one- and two-span bridges even though, in some cases, a slightly greater jacking force may be required. No difficulties should be expected for two-span structures with spans up to 200 ft (60 m) and with an unbalance of span lengths with a ratio of 1:2. Stressing should be done from the longer span of unbalanced continuous structures.

When a prestressing tendon changes direction, a resisting force is required to prevent the tendon from straightening. Although these forces are considered in the girder design, they are easily overlooked in the horizontal plane. This is especially true near the ends, where it is required to flare the girder stems in order to accommodate the anchorages. The situation should be reviewed if the system of prestressing used is different than what was assumed in the original design. The forces can be resisted by additional reinforcing steel or reduced by increasing the length of the flare, which reduces the angle change and amount of the transverse force.

Prestressed concrete box girder bridges are most commonly posttensioned by means of tendons placed in ducts embedded in the girder stems. In some cases, especially single spans, some of the tendons may also be placed in the soffit slab. After the tendons are tensioned, the ducts are filled with grout under pressure. The hardened grout protects the tendons and bonds them to the ducts to make them act integrally with the concrete structure.

Girders with bonded tendons, when subjected to overloads, develop very fine cracks at close intervals. These cracks close when the overload is removed. Even after loaded to ultimate, cracks adjacent to the ruptured section will close completely.

Girders with unbonded tendons, when subjected to overloading, develop widely spaced large cracks that do not close when the overload is removed. Reinforcing steel can be placed in girders with unbonded tendons in order to reduce the size and spacing of cracks caused by overloading.

If an unbonded tendon breaks, it becomes ineffective for the entire length of the tendon. If a bonded tendon should break, for any reason, it becomes ineffective for only a short distance—which may or may not be serious. The diameter of a tendon becomes smaller as it is tensioned. If a tensioned tendon that is embedded in hardened concrete or grout is broken, the broken ends tend to separate and withdraw into the hole that it cast in the concrete. The broken ends of the tendon have zero axial stress, and the tendon tries to return to its normal diameter. This action causes high radial stresses, which greatly increases the friction or effective bond. The broken ends thus become self-anchoring, and the full tension remains in the tendon a short distance from the break. Thus if a break occurs in an area of negative moment it may remain fully effective for positive moment, and vice versa.

A number of bridges have been built by placing the tendons inside the girder cells where they are not bonded to the structure. This method permits the use of much thinner girder stems but has the disadvantage that the structure has a reduced ultimate load capacity, and the protection of the tendons against corrosion must be considered.

Some systems of segmental construction require that tendons be placed in the ducts during early stages of construction and not tensioned until months later. Moisture, together with oxygen in the ducts, can cause considerable rusting of the tendons before they are grouted.

Difficulties have been experienced with prestressing ducts blocked by grout that leaked from adjacent ducts in which tendons were stressed and grouted in earlier stages.

Some structures have been constructed with the tendons placed

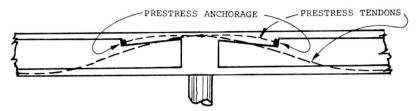

Fig. 16.1—Intermediate anchorage (elevation) in thickened portion of the deck slab.

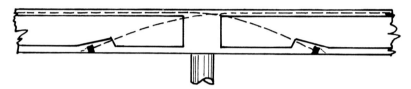

Fig. 16.2—Intermediate anchorage (elevation) in girder soffit.

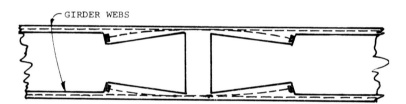

Fig. 16.3—Intermediate anchorage (plan view) in girder stems.

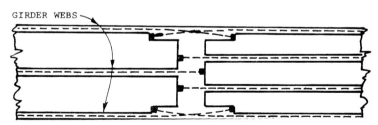

Fig. 16.4—Intermediate anchorage (in plan) accomplished by offsetting girder stems at bent cap or special intermediate diaphragm.

at the faces of the girders and the tendons encased in concrete after they were tensioned. Reinforcing steel extending from the girder into the encasing concrete increases the ultimate load carrying capacity of the structure and permits thinner girder stems. This scheme is particularly useful in unusual types of construction and when the structure is subjected to alternating positive and negative moments during construction.

For small bridges with only a few spans all the prestressing tendons ordinarily extend the full length of the bridge. Although all of the prestressing tendons may not be used efficiently under these conditions, the overall cost is generally less than using intermediate anchorages because of the relative cost of tendons, anchorages, and making provisions for tensioning the tendons at intermediate locations. In structures that are very long or large it is often economical to stop either some or all of the tendons at intermediate locations. This can be done by recessing the anchorages in thickened portions of the deck slab (Figure 16.1), soffit (Figure 16.2), and stems (Figure 16.3) or offsetting the girder stems at a bent cap or special intermediate diaphragm (Figure 16.4). If tensioning of the tendons is done from inside of the girder cells, access openings can be provided through the deck, soffit, or diaphragms. Access to adjacent cells can be obtained through openings in the girder webs. Since there is no need to gain access to the anchorages after the tendons have been tensioned and grouted, the access openings can either be closed permanently or provided with a removable cover. It is usually preferable to plug a deck opening permanently to avoid both a bump in the riding surface and future maintenance problems.

17
Lightweight Aggregate Concrete

Lightweight aggregate concrete has generally been used where dead load is an important factor and normal weight concrete is too heavy to be practical. It is a very suitable material for box girder bridge construction and in many areas is economically competitive with normal weight concrete. Since the physical properties of lightweight and normal weight aggregates are different, their design factors also vary. However, the design procedures are identical.

Lightweight aggregate concrete has been particularly useful in multilevel structures, where minimum structure depths are required and locations for columns are limited, and in very high bridges where the dead load of the superstructure requires unduly large columns and footings for seismic forces. The reduced weight of the concrete minimizes the amount of reinforcing steel in the superstructure and reinforcing steel and concrete in the substructure to the extent that the savings in materials can offset the slightly higher cost of lightweight aggregate.

In addition to many other types of structures, high-strength, lightweight aggregate concrete was used for constructing in Germany a 300-ft-span (90-m) pedestrian bridge by the free-cantilever method.

The results were excellent, and the lightweight aggregate concrete was credited for reducing the construction costs.[21]

The dead load stresses in a 750-ft-span (230-m) box girder cantilever bridge are about 90 percent of the total stresses. It is thus obvious that reducing the dead load is a logical approach for the construction of more economical longer spans.

At one time it was a popular belief of some engineers that lightweight aggregate concrete was not durable for the riding surface of bridge decks. Normal sand mortar toppings 1/2–1 in. (13–25 mm) thick were used as wearing surfaces on top of the lightweight aggregate concrete. The toppings were troublesome to place and, in a number of instances, spalled off. Experience and tests have proven that lightweight aggregates do give satisfactory and durable riding surfaces. In fact, lightweight aggregate concretes often provide better skid resistance than normal weight concretes. Tire chains and studded tires cause more wear on lightweight decks than normal concrete decks. However, in areas where tire chains are used, salt is commonly used for deicing. Because of the damage that salt may do to any type of bridge deck, it is the present trend to use additional protection to keep the salt from contacting the concrete surface. This additional protection should also safeguard it from any possible tire chain or studded tire damage.

The *1973 AASHTO Design Specifications for Highway Bridges* includes specifications for the use of lightweight aggregate concrete bridges.

A thorough discussion of the use of lightweight aggregate concrete for bridge construction is in a report by ACI Committee 213.[22]

18
Skewed and Curved Bridges

18.1 SKEWED BRIDGES

At present there is no universally accepted method for designing skewed box girder bridges. They are usually designed as square bridges with minor modifications. Research programs are being conducted that should increase the knowledge available and provide better methods for designing skewed bridges in the near future.

The effective design span for a skewed bridge is less than the centerline distance between abutments and piers. A skewed bridge that is designed as a square bridge without any modifications will be overdesigned in flexure and underdesigned in shear. For a structure that is very wide and has relatively short spans, the direction of bending is the perpendicular line between the supports. On a skewed structure that is quite narrow with relatively long spans, the longitudinal portion of a box girder superstructure that intersects a skewed bent acts, to some extent, as a large bent cap, which effectively reduces the assumed design span. If the same negative girder reinforcement that is determined from the moment envelopes is used for all of the girders, transverse deck cracking will be experienced in the acute

73

corners of the deck at the bents. Although noticeable, this cracking has never been known to be of any serious consequence.

Skewed bridges that have been designed as square bridges have developed cracks in the stems of exterior and interior girders at the obtuse and acute corners at continuous and simple supports.

There is a lack of published information concerning the design of skewed box girder bridges. The following interim guidelines have been formulated over a period of years based on reasoning and the observation of cracks in completed bridges:

1. The design shears for a skewed exterior girder should be multiplied by the secant of the skew angle. The result obtained by this procedure may obviously be out of line for extremely large skews, and judgment should be used to make suitable adjustments.

2. For bridges with fewer than six girders, none of the interior girders need be modified.

3. For bridges with six or more girders, the first interior girders should be designed for shear not less than the exterior girder shear modified for skew.

4. Girder flare lengths, stem thickness, and stirrup spacings should be adjusted logically and as repetitively as possible.

It is anticipated that better guidelines will be available in the near future; these will give satisfactory results more economically.

When a cap is entirely confined to the limits of the box girder superstructure, it can be oriented on a radial or normal to the girders at a single column bent without affecting clearances under the structure. It is thus possible for a bridge to be skewed with the roadways or waterway underneath but with the bridge itself designed and constructed as a nonskewed structure.

18.2 CURVED BRIDGES

Concrete box girder bridges are exceedingly adaptable to construction on the curved alignments which are required in modern highway construction. The box cross section is rigid torsionally, and torsional shearing stresses are comparatively small even though torsional moments creating them may be relatively large. Curvatures on main-line highway structures are generally so slight that they can be ignored completely for designing box girder bridges. More sharply curved structures on ramps and low-speed roads can often be designed as tangent structures, if the spacing of bents is not too great.

The effects of torsion in a box girder bridge are influenced by the radius of curvature; span lengths; out-to-out width of box structure; depth of structure; and thicknesses of deck, soffit, and exterior girder webs. For most curved highway structures, torsional shearing stresses can be calculated for the exterior girder stems and added to the dead load, live load, and impact shearing stresses.[23] Torsional shear stresses flow around the sides of each girder cell. These stresses generally cancel each other in the interior girder stems. There is no simple rule of thumb to determine whether torsion is a significant factor to be considered in the design of any particular curved structure. Approximate design methods are generally suitable for most highway box girder bridges. Exact design methods for curved structures are very laborious and are seldom used. Computers are very useful for the design analysis of very unusual structures when approximate design procedures might be questionable. This is generally true for pedestrian structures that have extreme curvatures. A designer develops a feeling for how significant the degree of curvature is for any particular structure after making a few design calculations.

19
Railroad Bridges

Reinforced and prestressed concrete box girder bridges are suitable for railroads as well as highways. Railroad structures carry heavier live loads than highway structures and by necessity are considerably more stocky. For Cooper's E-72 railroad loadings, the structure depth-to-span ratio is about 50 percent greater than for an HS 20–44 highway loading. The decks and girder stems are also considerably thicker.

The procedures for designing railroad and highway bridges are identical except for the design live and impact loads. The weight of ballast, if used, tends to make railway structures even heavier. Somewhat different details are naturally required to accommodate the railroad roadbed or other supports for the rails. In past years, refrigerator cars dripped salt brine on the roadbeds, which was injurious to bridge structures. More recently, however, modern mechanically refrigerated cars have eliminated this hazard.

The design and construction of conventional and unusual concrete box girder railroad bridges has kept pace with highway bridge construction in many areas. A prestressed concrete box girder railroad bridge with 184-ft (56-m) spans was built by segmental construction across the Rhone River in France in 1955.[24] Although railroads are generally carried on the top deck of a box girder bridge, some structures have been built that carry highway traffic on the top deck and railway traffic through the girder cells.[3,4,25]

A 1591-ft-long (495-m) bridge, with 377-ft (115-m) maximum spans and a top deck 130 ft (40 m) above the ground, was built in the city of Prague in 1968. The top deck accommodates six lanes of highway traffic, and a two-track rail line is carried through the 32 1/2-ft-wide by 17 1/2-ft-high (9.9 × 5.3 m) single cell. The main spans were built by the cantilever method, and the cast-in-place segments varied from 6 1/2 to 11 1/2 feet (2 to 3.5 m) in length.[25]

Ballast is ordinarily carried across a railway bridge for keeping

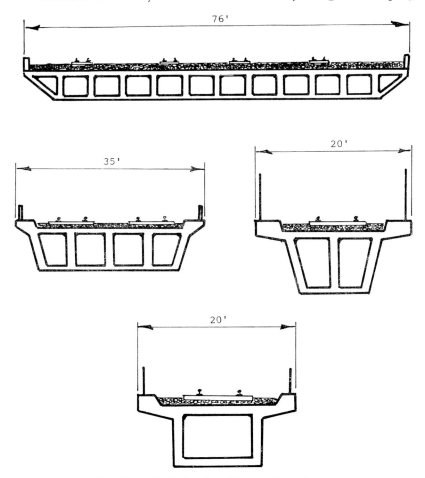

Fig. 19.1—Typical railroad box girder bridges.

the track in proper alignment and grade. The concrete deck should have drains or sufficient slope so that water does not become trapped. When the track grade is nearly level, the concrete deck can be sloped to provide drainage and the depth of ballast varied to meet the profile grade of the tracks. If the deck grades are nearly level, errors in determining camber and dead load deflection of a span can result in low spots that cause ponding of water either initially or after ultimate deflections occur.

In some instances it is not necessary to use ties and ballast, and the track can be set directly on pads and shims attached to the bridge deck.

Deck troughs confining ballast are usually waterproofed. There are a number of waterproofing methods; one system that works satisfactorily consists of attaching a 1/16-in. (0.63-mm) butyl rubber membrane to the deck surface and covering it with two layers of 3/8-in.-thick (10-mm) protective cover asphalt panel boards. The cover panels protect the rubber membrane from being punctured by the rock ballast.

Figure 19.1 shows the typical sections of a number of railroad bridges that have been constructed in recent years.

20
Segmental Prestressed

20.1 GENERAL

Most concrete box girder bridges have been built by casting them in place on falsework. In many instances, however, units have been cast as a complete span or as segments of a span some distance from the bridge site and then erected or assembled in their final position. This is usually necessary or desirable when the use of falsework is either impossible, undesirable, or uneconomical. However, segmental construction has also been used with falsework varying from a single temporary support at midspan to the same amount of falsework that would be required to construct the entire bridge in place.

Of all the different methods of assembling box girder bridge segments, the cantilever method has had the most influence on box girder bridge construction. The cantilever principle of construction was used as early as 1930 for a reinforced concrete girder type bridge with a 224-ft main span over the Rio do Peixe in southern Brazil.[26] The first segmental prestressed concrete box girder bridge constructed by the cantilever method was built in Germany in 1950.[27] This method basically consists of placing a segment (either precast or cast-in-place) on each side of a pier and stressing tendons that go through the segments and over the pier. This procedure is repeated with the segments on both sides of the pier balancing each other until

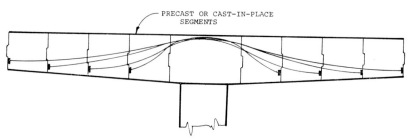

Fig. 20.1—Typical prestressing tendon arrangement for cantilever construction.

the cantilevers from adjacent piers meet at midspan (Figure 20.1).

The cantilever method of erection has been used in at least 24 countries throughout the world and has more than doubled the economical span length of box girder bridges. It permits the use of concrete box girders in locations where they would not be practical otherwise.

Segmental construction can be used for horizontally curved alignments, hump or sag vertical curves, and transitioned cross slopes. Single units can be made the full roadway width or, if the structure is very wide, with two or more lines of box segments constructed parallel or concentric with each other and joined with precast slabs or longitudinal closure concreting.

Segmental prestressed bridges can be constructed with commonly available equipment if the bridge is not excessively high and the space under the bridge is accessible (Figures 20.2 and 20.3). One of the main advantages of this type of construction is that it can be used over deep water, canyons, or even portions of cities where falsework or access from underneath is difficult or impossible. Segmental spans can be erected by working from the ground or from the top of the structure itself as construction proceeds.

Segmental prestressed construction has the advantage over other types of construction in that it can be done by a relatively small work force that performs repetitive tasks. Because of this, there is better control over workmanship and materials and, if all of the precasting can be done in one location rather than in place, greater economy of production can be achieved. This type of construction also permits a bridge to be erected in much less time, which can be important if it interferes with traffic or other activities. Segmental prestressed concrete box girder construction is adaptable to almost any situation. Segments 3 1/2 ft (1 m) long were used for a single lane bridge with a 131-ft (40-m) span in a primitive location with unskilled

Fig. 20.2—Precast segment being put into position for a highway project in Indiana (photograph courtesy of Portland Cement Association).

labor, limited construction time, limited funds, and shortages of conventional construction equipment.[28] Expensive custom-built equipment is not unusual for very large projects where the saving in time and labor more than pays for the great initial expense.

An end span can be stopped short and anchored in order to balance the cantilever in the adjacent span. Temporary intermediate bents can be used for temporarily supporting and stabilizing the uncompleted structure. Temporary towers can be placed at the piers or abutments with stays to temporarily support the cantilevers and control vibration. Other variations can be made to suit individual conditions. Horizontal and vertical guys have also been used to give longer cantilevers temporary support.

Fig. 20.3—Precast segments were used on the Stirling Bridge, Fremantle, W.A., Australia (photograph courtesy of Concrete Institution of Australia).

Box girder segments have been used in connection with almost every type of bridge construction including simple and continuous girder spans, cantilevers, arches, stayed girders, and suspension.

Segments can be joined and prestressed for special erection procedures and, after erection, some tendons removed or other tendons added to provide for service loading conditions. Supports of continuous structures can be raised or lowered to adjust moments to any desired condition. Many ingenious procedures have been used, and the possibilities are limited only by the imagination of the designer.

Box segments have ranged in length from 3 1/2 to over 50 ft (1–15 m). They have been cast in place, and they have been precast at the site and at yards many miles from the site. They have been conventionally reinforced and prestressed in either one, two, or three directions. They have been combined with other types of reinforced concrete, prestressed concrete, and steel construction.

One unique method of construction originated in Germany, also used in other countries, consists of casting segments of the box superstructure in lengths from 33 to 82 ft (10–25 m) long on the approach roadway behind the abutment. As each section is completed, it and the previously completed sections are jacked ahead onto the already completed piers in the river. A steel extension is attached to the first unit to assist in reaching the piers. Temporary piers were used between the permanent piers on some jobs, but a longer steel extension was used to eliminate the temporary piers on other bridges. Straight prestressing tendons are placed in the deck and soffit slabs to resist the longitudinal positive and negative bending moments as the structure is launched. Parabolic prestressing tendons are attached to the faces of the girder stems after the bridge has reached its final position. Since the girder forms stay in the same place, the concreting is quite similar to a factory operation.[3,4]

Superstructure costs in Europe have been estimated to be as much as 20 percent less than for structures built on falsework.

20.2 PRECAST SEGMENTS

The use of precast segments has a number of advantages over the cast-in-place method. In some cases very little special equipment is required, and the cost can be relatively small. Precast segments can be cast under controlled conditions independently of the substructure work. The use of a casting yard lends itself to better control, and if a proper production schedule is maintained an occasional breakdown will not affect the total construction time required to complete the job. When the substructure is completed, the segments can be erected and the superstructure completed in much less time than by casting the segments in place.

Although the lengths of precast segments vary considerably, they generally vary from 8 to 15 ft (2.4–4.6 m). The maximum practical size is limited by handling problems and hoisting equipment. The segment lengths in some structures have varied in order to keep the handling weights approximately equal. Holes for the prestressing

Fig. 20.4—Closeup of a typical precast segment shows keys for alignment and holes for prestressing tendons (photograph courtesy of Portland Cement Association).

tendons are cast in the segments (Figure 20.4). Many structures have been built with the tendons anchored in the girder stems, but there is a trend to locate the anchorages in haunches inside of the box section. This allows the placing of the segments and the stressing to be two independent operations.[29]

The segments can be cast end to end, in the same order that they will be assembled in the structure, on a bed that is at least as long as the span. The inner and outer forms are moved along the casting bed, which assures proper horizontal and vertical alignment. This method requires a considerable amount of space. The segments can also all be cast in a single location and each segment moved after it is completed. The segment is moved lengthwise, and the next unit is cast directly against the end of the previously completed unit. The ends of the segments are coated with a bond-breaking compound to facilitate separation. Segments for some bridges have been cast on end rather than in the normal horizontal position. Ends of the units are almost always keyed. Adjustments can be made in the forms for each unit to provide the correct horizontal and vertical alignment and the proper amount of warping for superelevation transition.[30,31,32]

One segmental box girder bridge was erected by placing all of the segments on timber falsework, concreting the gaps between the segments, placing the tendons in the box girder cells, and stressing from both ends.[33] Another was erected by setting the precast segments on two steel girders, longer than the bridge spans, so that the segments were in their final position. Four-inch (100-mm) gaps between the segments were concreted, tendons threaded through the ducts, and the entire span was posttensioned.[30]

Large steel trusses have also been used for supporting precast segments until they were stressed and self-supported. Trusses have been set below, alongside, and above the bridge superstructures on various jobs to accommodate clearance requirements.

Precast elements have been assembled on shore, joints concreted, the span tensioned, and the entire span floated into position on barges.[30]

Temporary ties attached to the deck and soffit slab have been used for holding two or more segments together in a number of different erection schemes. They permit the use of less erecting equipment or falsework.

Special care must be taken to line up the prestressing ducts at the joints. Slight offsets can make the placing of tendons difficult or even impossible. The ducts must be extended through the joints in such a manner that grout will not get into the ducts when the joints are concreted.

Depending upon conditions at the site, precast units can be erected in various ways. They can be lifted into place with a crane either on the ground or a barge. They can be lifted off a barge or a truck under the structure by means of lifting equipment on the deck of the previously erected units, or they can be delivered and placed by means of a launching gantry supported by the completed bridge piers and/or completed portions of the superstructure.

Figure 20.5 illustrates the operation of a particular type of gantry. Figure 20.5(a) shows the gantry being supported by two legs that are set on the previously completed span and cantilever of the bridge. The precast deck unit that rests directly on top of the next pier has been transported across the completed portion of the bridge, carried by a hoist that runs on rails attached to the lower chord of the gantry, and is being lowered to its final position on top of the next pier. In Figure 20.5(b) a temporary extension has been placed on top of the next pier, which supports the gantry so that it can be

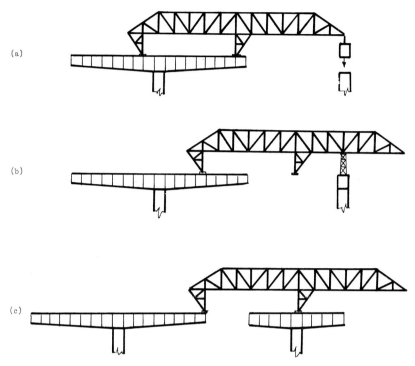

Fig. 20.5—Typical use of launching gantry for cantilever construction.

moved ahead to its new position shown in Figure 20.5(c). Eight precast segments have been delivered across the completed portion of the
bridge, and the gantry has placed them in pairs on each side of the
new pier. When both cantilevers on this pier have been completed,
the two cantilevers will be joined in the center span and the gantry
will be moved ahead to repeat the cycle.

Launching gantries are frequently used when the area under the
bridge cannot be used for delivering precast units or operating erection equipment. Gantries are very expensive and are only economically suited for very large projects. The first launching gantries were
constructed for specific bridges and used only once. For the sake of
economy, later ones were made demountable with higher strength
steels and bolted joints so they could be dismantled and used for other
jobs. Some of the more recent ones were made lighter and more
portable by using triangular cross sections. Groups of cable stays are
sometimes used to reduce deflections when transporting and placing
segments. The gantries are generally much narrower than the seg-

ments they handle, and the segments must be rotated 90 deg to enable passage through the legs of the gantry. If the bridge being constructed has sharp curves and superelevations, the gantry must maintain stability while placing the segments accurately. Gantries can be used in a wide variety of methods.

Although the precast units of some structures have been butted together with dry joints, the adjoining faces are generally coated with epoxy, cement grout, or concrete. Epoxy coatings are usually less than 1/32 in. (1 mm) thick and require accurate matching surfaces—usually obtained by casting a segment directly against the adjacent segment treated with a bond-breaking agent. Epoxy acts as a very efficient lubricant when placing the segments; is a glue that assists in making the segments act as a single homogeneous unit; helps to even out minor irregularities between the mating surfaces; and waterproofs the joints.

Tests of thin epoxy resin joints, made in a number of different countries, indicate it is possible to develop 94 percent of the flexural tensile strength and approximately 75 percent of the shear strength of a comparative monolithically cast specimen.[30] The shear keys assist in keeping the mating surfaces locked together until the epoxy has set.

Special care must be taken when manufacturing precast segments that will be butted end to end with not more than a coat of epoxy between them. When a series of adjacent segments are cast in a single form and each segment is moved forward to act as a form for the next one, it is inevitable that the two segments will not maintain their proper geometric relationship. When the concrete has set, the segments should be carefully surveyed before they are separated and proper compensations should be made in casting the subsequent segments. If suitable corrections are not made, it may be difficult to make the cantilevers meet at midspan or to obtain a satisfactory alignment of the roadway.

Mortar or unreinforced concrete joints ranging from 1 to 4 in. (24–100 mm) have been used frequently, and widths up to 16 in. (400 mm) have been used. Reinforced concrete has been used in joints from 8 to 24 in. (200–600 mm) wide. Reinforcing steel from adjacent segments is usually lapped or welded in the wider joints.[30] Thick mortar and concrete joints do not require the exacting fit between the ends of the units that is necessary for epoxy joints, but they are not as well suited to cantilever type construction. The wider joints are better suited for bridges constructed on falsework.

Joints between precast units are generally prominent because of the epoxy or grout between the units. Since there is generally no reinforcing steel through the joints, design and erection stresses must be calculated very carefully. In a number of instances joints between precast segments have opened up because of extreme and rapid changes in temperature. Errors in camber are difficult to diagnose and correct. Prestressing systems are limited to ones in which the tendons can be threaded through the ducts.

Cantilever construction using precast segments is very fast, and erection speeds of 40 ft (12 m) per day are common.[30] For large structures precast segmental construction can be much faster than for cast-in-place segmental and other types of construction.

20.3 CAST-IN-PLACE SEGMENTS

Basically, the use of cast-in-place segments for cantilever construction consists of casting a short longitudinal segment of the superstructure on each side of a pier and tensioning prestressing tendons through the segments as soon as the concrete has attained sufficient strength. The forms are then moved outward, new segments cast against the ends of the previously completed segments, and the same procedure is repeated over and over until the cantilevers reach midspan and are joined with the cantilevers from the adjacent piers.

Although there has been considerable variation in lengths of cast-in-place segments, the most commonly used lengths for cantilever construction have been from 8 to 16 ft (2.4–4.9 m). Some structures have larger segments next to the piers because the greater weights can be supported directly by the pier. The segments generally attain sufficient strength for stressing between 30 and 60 hours after casting. All segmentally constructed spans over 400 ft (120 m) have been built with cast-in-place segments. Pairs of segments are commonly completed in one-week cycles, but two pairs per week have been realized. The rate of progress usually averages from 3 to 6 ft (1 to 2 m) per day. This type of construction requires at least a pair of specially designed travellers, which may cost from $50,000 to $200,000 and possibly cannot be used again on other jobs.

The construction joints between the cast-in-place segments are not as prominent as the joints between precast segments, and the reinforcing steel is continuous through the joints. One of the advantages of casting the segments in place is that if actual cambers do not agree with predicted cambers, as construction progresses, corrections can be made by revising the alignment of following segments. Since the tendons can be in place when the segments are cast, almost any type of prestressing tendons can be used—bars with couplers, button-headed wires, strands, and so on.

Fig. 20.6—Pine Valley bridge, California.

The Pine Valley bridge, California, has a maximum span of 450 ft (137 m) and was constructed by casting the segments in place (Figure 20.6).

20.4 DESIGN

Although most bridge designers are aware of the phenomenon of shear lag, the only attempt to account for it in design specifications is to limit the assumed effective width of compressive flanges. These limits are based on the span length of the girder, the girder spacing, and the thickness of the slab. Although these limits are quite satisfactory for ordinary highway separation structures and river bridges, they should be questioned for the design of unusual or monumental type bridges. Mathematical analysis and models have been used for predicting shear lag in unusual bridges.

Prestressing, dead load, concentrated live loads, and uniform live loads each have different shear lag effects in a bridge. Shear lag for each type of loading is also affected by span length, structure width, girder spacing, deck cantilever length, structure depth, thickness of deck, and thickness of soffit. The problem becomes even more complicated in cantilever bridges where each short segment is prestressed to a different degree.[19]

Although segmental box girder bridges have been designed by the ordinary concrete beam theory, analysis by the folded plate, thin walled beam, or finite element methods[30] are commonly used. The latter three methods are usually used because of doubts concerning shear lag and the nonlinearity of stress variation. To date, very little test data is available on full scale bridges. In one project the finite element analysis gave about 50 percent variation in stress across the bottom flange, but preliminary test results indicated a stress variation of less than 10 percent. The present methods of analysis do not give full consideration to the fact that the structures are erected in segments with many stages of longitudinal and transverse prestressing.

The Portland Cement Association constructed and did extensive testing of a 1/10-scale microconcrete model of a proposed bridge across the Potomac River. The prototype bridge on which the model was based has a 750-ft (229-m) main span with a 440-ft (134-m) span on each side. The roadway deck is 110 ft (34 m) wide, and the side spans are curved in plan. The exterior girders are sharply curved in the vertical direction. Testing of a structural model was considered desirable since the proposed bridge would set several precedents. Application of a service load, representing dead load of the prototype and live load plus impact equal to an HS 20–44 loading, did not produce any cracking, and the structure remained essentially elastic. The model carried an extreme overload of $1.5D + 2.5(L+1)$ with only minor cracking. The cracks caused by overloading closed until they were barely visible to the naked eye when the overload was removed and the condition of $1.0D$ was restored.[34]

Most segmental prestressed structures have been designed and built by designer-contractor companies and have been adapted to one particular type of operation and prestressing.

The weight of construction equipment and all other factors must be considered in the design of the structure for each stage of construction. The construction loads imposed on a segmental prestressed structure erected by the cantilever method are usually as great as any that will be placed on it during its service life. It is therefore pretested before it is opened to use. Most bridges are designed for an unbalance of one segment in each stage in order to provide for unplanned construction operations.

Roadway decks of segmental prestressed bridges are generally prestressed transversely. Prestressing permits the use of a thinner deck, which reduces the dead load of the segments, permits longer segments be used, and gives a superior deck. The reduced dead load also reflects other economies. Transverse shortening of the deck due to prestressing can induce substantial stresses in the girder stems if they are quite thick. This is usually most critical near the supports, where the stems are thickened to resist shear and the soffit slab is thickened for compression.

The thicknesses of all the various parts of a box are determined partly by stress and partly by clearances required for the tendons and anchorages. Girder stems are frequently prestressed either vertically or diagonally in order to avoid any tension in the concrete.

Spans up to about 300 ft (90 m) are usually built with a constant depth. Longer spans require excessively thick soffit slabs in order to furnish the required compressive area and it becomes advantageous to haunch the girders at the piers.

The end spans of a typical three-span bridge, constructed by the balanced cantilever method, should preferably be only 65 percent of the main center span.[29]

Some box girder cantilever bridges have been built with hinges joining the cantilevers at midspan. The hinges transmit vertical shear from one cantilever to the other, but no angular restraint. Shock absorbers have been used in some cases which resist instantaneous loads but adjust to sustained loads.[35] Connecting the ends of two cantilevers in a span by means of a hinge has a number of disadvantages. When the sun shines on the deck, it expands because of the heat and if the two cantilevers are joined by a hinge the center of the span may go down appreciably. This may produce an unsightly angle point at midspan as well as an undesirable riding surface. The hinge point will also go down over a period of time because of the weight and creep characteristics of the cantilevers.[36]

Many bridges have been designed and constructed with hinges at midspan, including one of the world's longest concrete box girder spans 755 ft (230 m), across Urato Bay, Japan. It has also been proposed to hinge the center of the 790-ft (241 m) span that is to be built over a channel between two islands in the Pacific Trust Territory. It is thus obvious that the disadvantages of midspan hinges noted in this section are factors that should be given serious consideration rather than arguments for not using them under any circumstances.

Most recent spans have been made continuous rather than hinged at midspan. This is done by placing a cast-in-place section between the ends of the cantilevers and tensioning continuity tendons. Continuity tendons are passed through the bottom of the closure section and anchored in either the soffit or the deck. Continuous spans are less sensitive to vertical and angular displacements at the ends of the cantilevers, the vertical alignment is smoother, the spans require less maintenance, provide better dynamic stability, and have a higher capacity for overloading.

The actual prestressing force has a much greater influence on the deflection of a cantilever span with a hinge at midspan than on a similar but continuous span. For one particular structure it was calculated that a 5 percent variation in prestressing force would increase the deflection of a span with a hinge at midspan by 23 percent. The same variation would affect the deflection of a continuous span by only 7 percent. In addition, the continuous span is three times as rigid as the hinged span under live load.[31] If a hinge is required in a span in order to accommodate changes in length of a bridge, it is preferable to place it near the point of inflection. This reduces the angular difference in grades on the two sides of the hinge. Placing the hinge away from the centerline of span generally requires a special construction procedure and should be planned so that the erection schedule will not be delayed.

It has been found by experience that closure placements joining adjacent cantilevers should be kept as small as possible and made at night. The variations in the distance between the ends of the cantilevers caused by changing temperatures in the daytime destroy the freshly placed concrete. In some cases it may be required to jack the cantilevers horizontally or vertically in order to adjust the moments before the permanent bearings or closure concrete is placed. Another method of adjusting moments is to construct the superstructure on temporary bearings at a pier and then raise or lower it to the proper elevation after the spans are completed.

Provisions must be made in all long structures for accommodating longitudinal variations in length. Expansion joints are normally required in box girder cantilever construction every 1000–1500 ft (300–460 m). These lengths may vary considerably depending on the heights and flexibility of the piers, joint details, and other factors.

Some designers prefer to construct the connection between the box girder superstructure and the tops of the piers so that no moments can be transferred from one to the other. Other designers have a tendency to prefer rigid connections.

Prestressing forces and creep in the concrete result in axial shortening of the cantilevers. The dead load of the cantilevers themselves, permanent loads added to the structure, and creep of the concrete and prestressing steel result in downward deflections of the spans. If the cantilevers are made continuous at midspan, redistribution of forces and moments due to these deformations should be considered in the design.[18]

20.5 DEFLECTION AND CAMBER

Unless compressive stresses are uniform throughout the depth of a cantilever section at all times and at all locations along its length, the end will deflect vertically. This uniform stress condition is impossible to achieve, and compressive stresses are always higher at the bottom. Prestress loss due to steel relaxation makes the situation worse, and concrete creep is also responsible for causing the end of the cantilever to sag. Considerable research and more thorough analysis concerning the properties of concrete and its reaction to applied loads at various ages and for various durations, increments, and fluctuations of loading could improve the prediction of initial and long-term deflections. The usual empirical rules that are used for assessing creep, stress losses, and deflections for ordinary prestressed structures are of little help in the design of segmentally constructed bridges.

The problem of controlling the deflections of cantilevered segmental bridges is considerably more complicated and difficult than for construction of spans supported by falsework.

If dead load deflections were ignored completely, cantilevers would sag and curve downward until they met at midspan. It is thus desirable to camber the segments upward so when they deflect downward they will ultimately be as close as possible to the planned profile grade.

The deformation and movement of each cast-in-place segment is dependent on the deformation and movement of the column and each preceding segment in the frame. Each time a new segment is completed, each preceding segment is a different age, has different creep characteristics, and is subjected to a different loading. In order to be assured that the cantilevers meet at midspan and the roadway has the proper profile grade, it is necessary to tabulate the effect of the new loading on each of the preceding segments each time an additional segment is added. Field measurements are taken as construction progresses, and corrections are made to the alignment of each new segment as required to obtain the correct final profile. If work is delayed for a time, or if any change affects the creep characteristics of the concrete segment, all previous calculations may become erroneous. The calculations involved are obviously extensive, but much labor can be saved by using electronic computers. The sun shining on the deck and faces of the column causes the ends of the cantilevers to move constantly. It is therefore necessary that all measurements used for determining corrective action be taken when all parts of the bridge are of nearly uniform temperature—or at least relatively the same.

Because precast segments are aged more than cast-in-place segments before being placed in the structure they have more dimensional stability. Concrete strains that result in deflection of the span are a function of the age of the concrete. For example, the ultimate creep strain of three-day-old concrete is 2.5 times greater than the ultimate creep strain of three-month-old concrete for the same applied load.[29]

The age of each segment when it is placed in the structure should be known in order to correctly predict its creep characteristics, which are required to calculate the deflection of the cantilever. It is usually very difficult to make corrections to the camber of a structure with precast segments. Corrections can be made most readily if closure sections are made between the precast units. In some cases corrections can be made by jacking at the bents or by adjusting the prestressing forces.

When two cantilevers meet at midspan, the ends are jacked to bring them into alignment before the closure section is placed. The amount of this jacking is generally so small that it is insignificant for design purposes.

21
Utilities

Box girder structures are particularly well adapted to carrying pipelines, conduits, cables, and other utilities. The large cells provide adequate space for numerous utilities that are kept completely out of view. Some box girder bridges have used the cells for carrying large amounts of drainage—either in large pipes or, in some cases, by using the girder cells as box culverts.

21.1 SAFETY

Consideration should be given to the consequences of a leak or rupture of a utility line carried inside a box girder. Gases and flammable liquids could cause serious explosions or fires. Oil leaks could cause accidents on roadways under the bridge or pollute waterways as well as creating fire hazards. A ruptured, large, high-pressure waterline might quickly fill the cells with water and cause structural damage. Liquids flowing from a ruptured line through soffit vents would be a hazard to traffic under the bridge. When any one of these accidents is considered possible, it is advisable to place the utility inside of a larger diameter pipe extended into the approach roadway fill and vented so that any escaping material will be carried to a safe location.

Fig. 21.1—Typical utility being placed in box girder cell during construction.

21.2 INSTALLING UTILITIES

Utilities are occasionally placed in completed box girder bridges by coring suitable holes through the abutment backwalls, diaphragms, and bent caps. Care must be taken to ensure that the structural integrity of the bridge is not compromised by removing critical portions of the structure and that the methods of doing the work are satisfactory.

Utilities are ordinarily placed in the box cells during construction (Figure 21.1). If the utilities are not already located at the bridge site, or if they can be rerouted during construction, there is usually very little difficulty in placing them in the cells during or after construction. It is occasionally necessary to construct a bridge which will carry an existing utility located at the bridge site when it is not feasible to reroute the utility or interrupt service. This has been done by a number of different methods, depending upon particular circum-

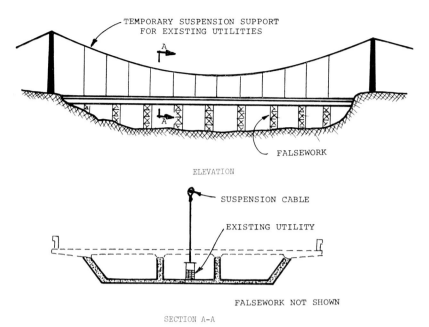

TEMPORARY SUSPENSION SUPPORT
FOR EXISTING UTILITIES

FALSEWORK

ELEVATION

SUSPENSION CABLE

EXISTING UTILITY

FALSEWORK NOT SHOWN

SECTION A-A

Fig. 21.2—Support for existing utility.

stances. In general, the utility must be protected and strengthened so that it can be supported for raising and lowering. One commonly used method of supporting an existing utility in place, so that the bridge can be built around it, is to use suspension cables that span the entire site (Figure 21.2). Another method is to drive piles or construct columns on either side of the utility that support cross beams from which the utility can be suspended (Figure 21.3). If the cross beams cannot span the entire width of the structure, temporary holes can be cast in the soffit around each support. For small utilities, the utility can be temporarily attached to or suspended alongside single piles or columns. The utility is lowered to its final position on the soffit after the soffit slab and stems are completed, the supporting structure removed, the holes in the soffit filled, and the structure completed in the usual manner.

Deck forms are ordinarily left in place. It is an expensive procedure to remove them through the relatively small openings that can be provided, and very little of the material is salvable. When access to the cells is desired, for maintenance or installing future utilities, either all of the forms should be removed or sufficient portions should

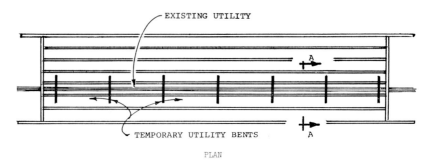

PLAN

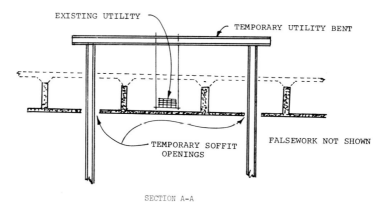

SECTION A-A

Fig. 21.3—Support for existing utility.

be taken out to provide necessary working space. Any material that remains in the cells should not interfere with or present a safety hazard to workmen who must enter the cells in the future. A minimum crawl space 2 ft (0.6 m) wide by 3 ft (0.9 m) high should be provided throughout the length of all such cells.

21.3 ACCESS TO UTILITIES

Access to the utilities in a box girder bridge may be had through manholes in the deck or special openings in the soffit slab. Covers should be considered for use on soffit openings in order to keep out animals and transients. In special cases, where frequent access is required and it is not desirable to go through the deck or soffit, a gallery can be constructed between the end diaphragm and abutment backwall with a door located next to one of the exterior girders. Some utilities are customarily placed inside of larger pipes so that the utility can be withdrawn or inserted from an excavation at the end of the bridge rather than making repairs or replacements inside of the cells.

21.4 FUTURE UTILITIES

It is quite common for utility companies to request permission to locate their utilities on a bridge after it has been completed. This is usually due to unanticipated growth of communities and the problems involved in placing utilities under rivers or heavily traveled highways.

It is undesirable, for aesthetic reasons, to hang utilities on the outside of a bridge. Provisions can generally be made for future utilities during initial construction of box girder bridges for relatively little expense. Openings through caps and diaphragms should be made in some location where the utilities can be installed with a minimum of interference to the traveled way—usually in the shoulder or median area.

Openings in abutment backwalls can be closed with thin concrete "knock out" panels, bricks and mortar, or some other material that can withstand the pressure of the earth behind the abutment but which can be easily removed at some future date.

References

1. Dallaire, G., "Designing Bridge Decks That Won't Deteriorate," *Civil Engineering—ASCE*, V. 43, No. 8, August 1973, pp. 43–48.

2. Sisdiya, R. G.; Ghali, Amin; and Cheung, Yau Kai, "Diaphragms in Single- and Double-Cell Box Girder Bridges with Varying Angle of Skew," ACI JOURNAL, *Proceedings* V. 69, No. 7, July 1972, pp. 415–419.

3. Leonhardt, Fritz, "Long Span Prestressed Concrete Bridges in Europe," *Journal*, Prestressed Concrete Institute, V. 10, No. 1, February 1965, pp. 62–75.

4. ———, "Advances in Bridge Design with Prestressed Concrete," *Design Seminar*, Prestressed Concrete Institute, Chicago, January 26–29, 1971, pp.13–1 to 13–35.

5. Campbell-Allen, Denison, and Wedgewood, Raymond J. L., "Need for Diaphragms in Concrete Box Girders," *Proceedings*, ASCE, V. 97, ST3, March 1971, pp. 825–842.

6. Gamble, W. L., "Ineffectiveness of Diaphragms in Prestressed Concrete Girder Bridges," *Bulletin* No. IHR–93, Department of Civil Engineering, University of Illinois, Urbana, April 1972, 12 pp.

7. Van Horn, D. A., "Proposed New Specification for Lateral Distribution of Loads for Bending Moment in Prestressed Concrete Spread Box Beam Bridges," Lehigh University, AASHO-PCI Joint Committee on Bridges and Structures, March 1972, 8 pp.

8. Hanson, Norman W., "Precast-Prestressed Concrete Bridges—2. Hor-

izontal Shear Connections," *Journal,* PCA Research and Development Laboratories, V. 2, No. 2, May 1960, pp. 38–58.

9. Carpenter, J. E.; Hanson, J. M.; Fiorato, A. E.; Russell, H. G.; Meinheit, D. V.; Rosenthal, I.; Corley, W. G.; and Hognestad, E., "Design of Bent Caps for Concrete Box Girder Bridges," *Research and Development Bulletin* No. RDO32.01E, Portland Cement Association, Skokie, January 1974, 24 pp.

10. Kriz, L. B., and Raths, C. H., "Connections in Precast Concrete Structures—Strength of Corbels," *Journal,* Prestressed Concrete Institute, V. 10, No. 1, February 1965, pp. 16–47.

11. Lenzen, K. H., "Bridge Vibrations—Results of Questionnaire to Bridge Engineers and Discussion by Author," Committee on Bridge Dynamics, Highway (Transportation) Research Board, Washington, D.C., February 1970, pp. 1–9.

12. Smith, A. W., "The Newmarket Viaduct. Part 2: Design Investigations and Specification," *New Zealand Engineering* (Wellington), V. 20, No. 12, December 1965, pp. 493–511.

13. "Temperature Changes Cause Bridge Failures," *Australian Civil Engineering,* March 1970, pp. 36–37.

14. "Aukland's Newmarket Viaduct—Its Current Behavior," *RRU Newsletter* No. 20, Road Research Unit, New Zealand National Roads Board, Wellington, July 1968, pp. 16–18.

15. "Progress in Rectifying Faults in the Newmarket Motorway Viaduct," *RRU Newsletter* No. 28, Road Research Unit, New Zealand National Roads Board, Wellington, July 1970, p. 8.

16. Leonhardt, F., and Lippoth, W., "Conclusions Drawn from Distress of Prestressed Concrete Bridges," *Beton und Stahlbetonbau* (Berlin), V. 65, No. 10, October 1970, pp. 231–244.

17. Stewart, C. F., "Annual Movement Study of Bridge Deck Expansion Joints," *R & D Report* No. 2–69, California Division of Highways, Sacramento, June 1969, 52 pp.

18. Janssen, H. H., "Design of Precast Segmental Bridges," Seminar, Prestressed Concrete Institute, Chicago, September 25, 1973, 28 pp.

19. Meggs, R. C., and Base, G. D., "Shear Lag in Box Girder Bridges," *Proceedings,* Sixth Conference of the Australian Road Research Board, Victoria, 1972, V. 6, Part 5, pp. 381–402.

20. Reynolds, Joseph C., and Emanual, Jack H., "Thermal Stresses and Movements in Bridges," *Proceedings,* ASCE, V. 100, ST1, January 1974, pp. 63–78.

21. Finsterwalder, Ulrich, "Free Cantilever Construction of Prestressed Concrete Bridges and Mushroom-Shaped Bridges," *First International Symposium on Concrete Bridge Design,* SP–23, American Concrete Institute, Detroit, 1969, pp. 467–494.

22. ACI Committee 213, "Guide for Structural Lightweight Aggregate Concrete," ACI JOURNAL, *Proceedings* V. 64, No. 8, August 1967, pp. 443–467.

23. Lovering, P. H., "Hollow Box Girder Viaduct Vital for Complex Traffic Interchange," *Modern Developments in Reinforced Concrete,* Portland Cement Association, Skokie, 1953, pp. 1–23.

24. "Railway Bridge over France's Rhone River Marks Bold Use of Prestressed Concrete," *Engineering News-Record,* V. 159, August 1, 1957, pp. 36–38.

25. "Viaduct Serves as Road, Rail Bridge," *Engineering News-Record,* V. 181, December 5, 1968, pp. 22–23.

26. Schjodt, R., "Long Rigid Frame Bridge Erected by Cantilever Method," *Engineering News-Record,* V. 108, August 6, 1931, pp. 208–209.

27. Finsterwalder, Ulrich, "Prestressed Concrete Bridge Construction," ACI JOURNAL, *Proceedings* V. 62, No. 9, September 1965, pp. 1037–1046.

28. "Ingenuity and Prestressing Make a Low Cost Bridge," *Engineering News-Record,* V. 174, April 1, 1965, pp. 30–32.

29. Muller, Jean, "Ten Years of Experience in Precast Segmental Construction," *Proceedings,* 42nd Annual Convention, Structural Engineers Association of California, San Francisco, 1973, pp. 16–38.

30. Lacey, Geoffrey C.; Breen, John E.; and Burns, Ned H., "State of the Art for Long Span Prestressed Concrete Bridges of Segmental Construction," *Journal,* Prestressed Concrete Institute, V. 16, No. 5, September–October 1971, pp. 53–77.

31. Muller, Jean, "Long Span Precast Prestressed Concrete Bridges Built in Cantilever," *First International Symposium on Concrete Bridge Design,* SP-23, American Concrete Institute, Detroit, 1969, pp. 705–740.

32. "Precast Segmental Bridge Structures," Brochure by Freyssinet International, Boulogne (France) 1973, 27 pp.

33. "Prestressed Girders Make Australian Bridge," *Engineering News-Record,* V. 172, February 3, 1964, pp. 94–95.

34. Corley, W. G.; Russell H. G.; Cardenas, A. E.; Hanson, J. M.; Carpenter, J. E.; Hanson, N. W.; Helgason, T.; and Hognestad, E., "Design Ultimate Load Test of 1/10-Scale Micro-Concrete Model of New Potomac River Crossing, I–266," *Journal,* Prestressed Concrete Institute, V. 16, No. 6, November–December 1971, pp. 70–84.

35. Gerwick, Ben C., Jr., "Bridge over the Eastern Scheldt," *Journal,* Prestressed Concrete Institute, V. 11, No. 1, January–February 1966, pp. 53–59.

36. Kokubu, Masatane, "Deflections of Prestressed Concrete Bridges in Japan," *Magazine of Concrete Research* (London), V. 24, No. 80, September 1972, pp. 117–126.

Index